Entangled Magazine

Published by Anthony Patch

July 2021

Table of Contents

Upon initial examination in January of 2020 of the genetic sequence of SARS-CoV-2, my wife Kathleen and myself arrived at significant conclusions regarding this virus, and any potential vaccines developed against it.

First, both the virus and all vaccines integrate foreign genetic material into the DNA of humans. We felt confident subsequent research literature eventually would confirm our expectations. Indeed, the peer-reviewed scientific white papers have confirmed this, as referenced within the pages of this magazine since January of 2020.

Second, unequivocally, vaccines specific to SARS-CoV-2 and its variants, are delivering the mechanisms manifesting as the Biblical mark of the beast. This is accomplished by permanently altering our DNA, thus, God's creation of us in his image. The human body was designed as the temple of the Holy Spirit. Genetic alteration, through the molecular biological actions of manmade genomic sequences of mRNA, both in the form of the virus and all subsequent vaccines, results in genetic (DNA/RNA) changes that God does not recognize as His original human design. We can conclude that the inspiration for these novel genomic sequences is satanic in origin. The primary goal is the damnation of souls following a free will decision to take the vaccine.

Revelation 13: 8

And all that dwell upon the earth shall worship him, whose names are not written in the book of life of the Lamb slain from the foundation of the world. KJV

Subsequent peer reviewed research literature has proven integration of human DNA by both the virus and the vaccines derived from it. The results were the same: chimeric DNA.

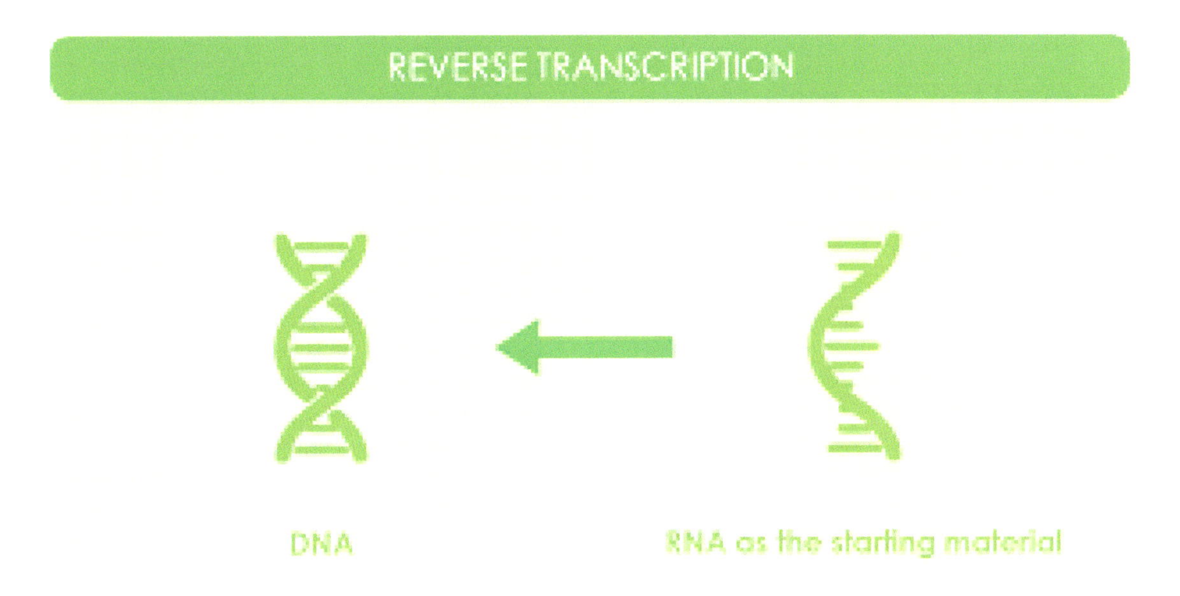

Researchers from MIT and Harvard published evidence that the SARS-CoV-2 RNA can, through the actions of the enzyme, transcriptase, reverse transcribe mRNA to DNA. Double-stranded, complimentary DNA then integrates by the actions of the enzyme, integrase, into human DNA. Their investigation was inspired by the many patients who continued testing positive for COVID-19 *after* the virus had already cleared from their body. The researchers discovered chimeric transcripts (this is the actual scientific term) containing viral DNA sequences integrated in patients who recovered from COVID-19.

Since COVID-19 often induces a cytokine storm in severe cases, researchers confirmed the possibility of enhanced reverse transcriptase activity through an *in vitro* study using cytokine-containing conditioned media in cell cultures. They found a 2-3-fold upregulation of endogenous (within a system) LINE-1 (long interspersed nuclear elements making up 20% of the human genome) expression in response to cytokines. The exogenous (originating from outside) RNA from the virus incorporated into human DNA could produce fragments of viral proteins indefinitely after the infection has been cleared. This yields a false positive on a PCR test.

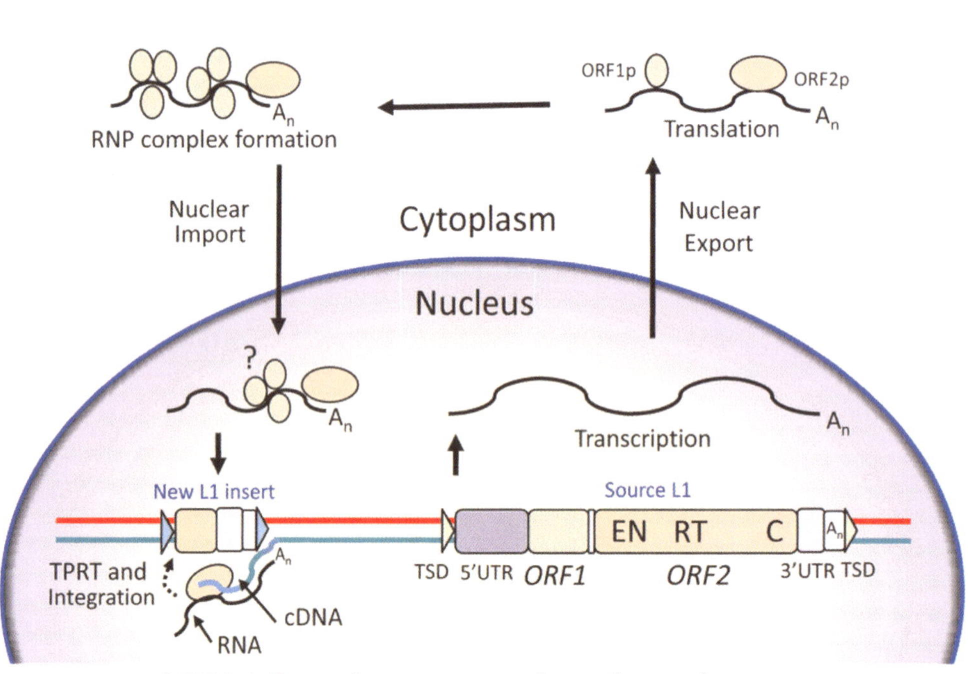

LINE-1 (long interspersed nuclear elements

The combination of fragments of viral proteins, referred to in the literature as VLPs (virus-like particles), and Spike (S) proteins within all vaccines for SARS-CoV-2 has resulted in the ongoing exponential increase in reported (and unreported) cases of severe side effects and deaths.

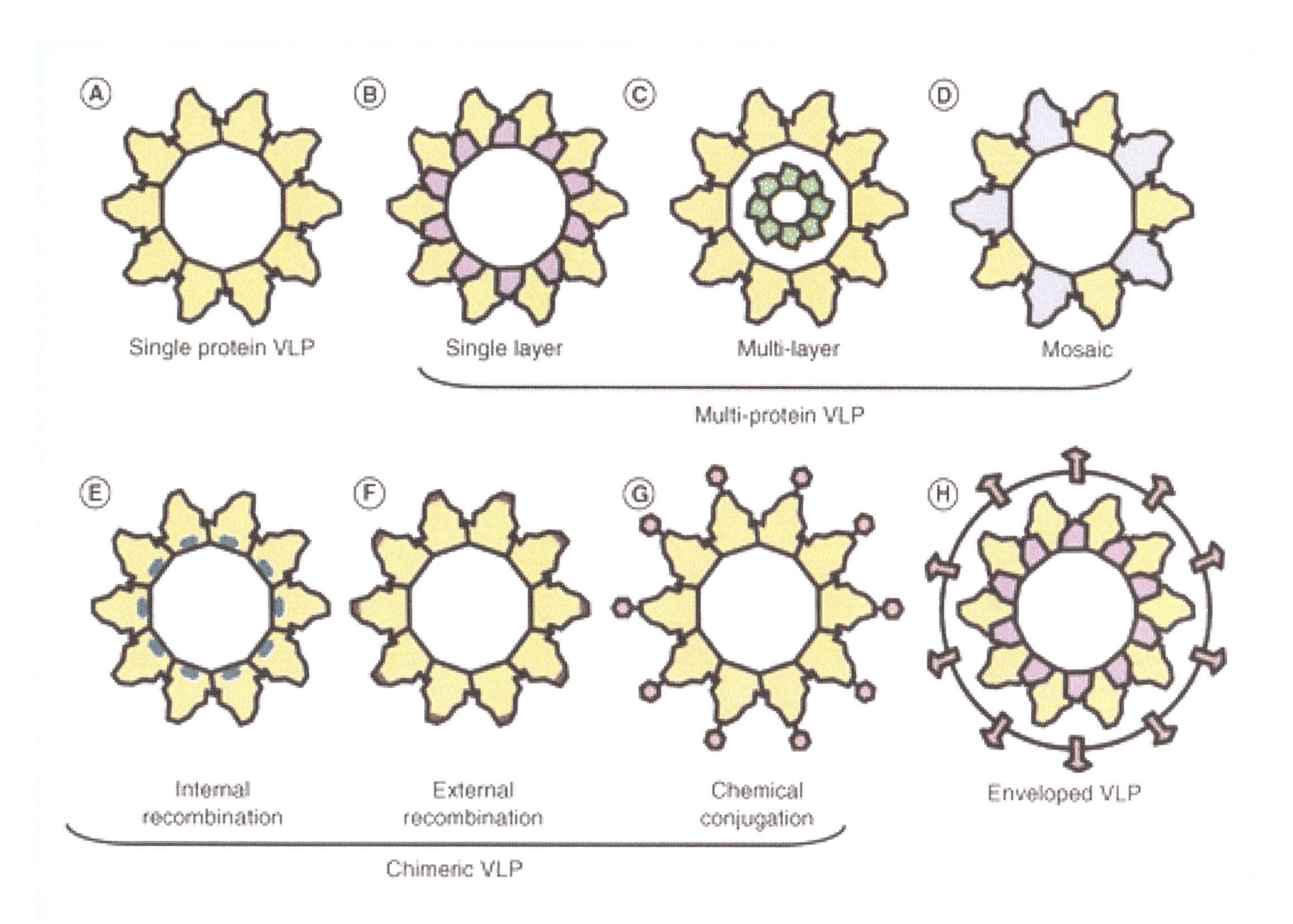

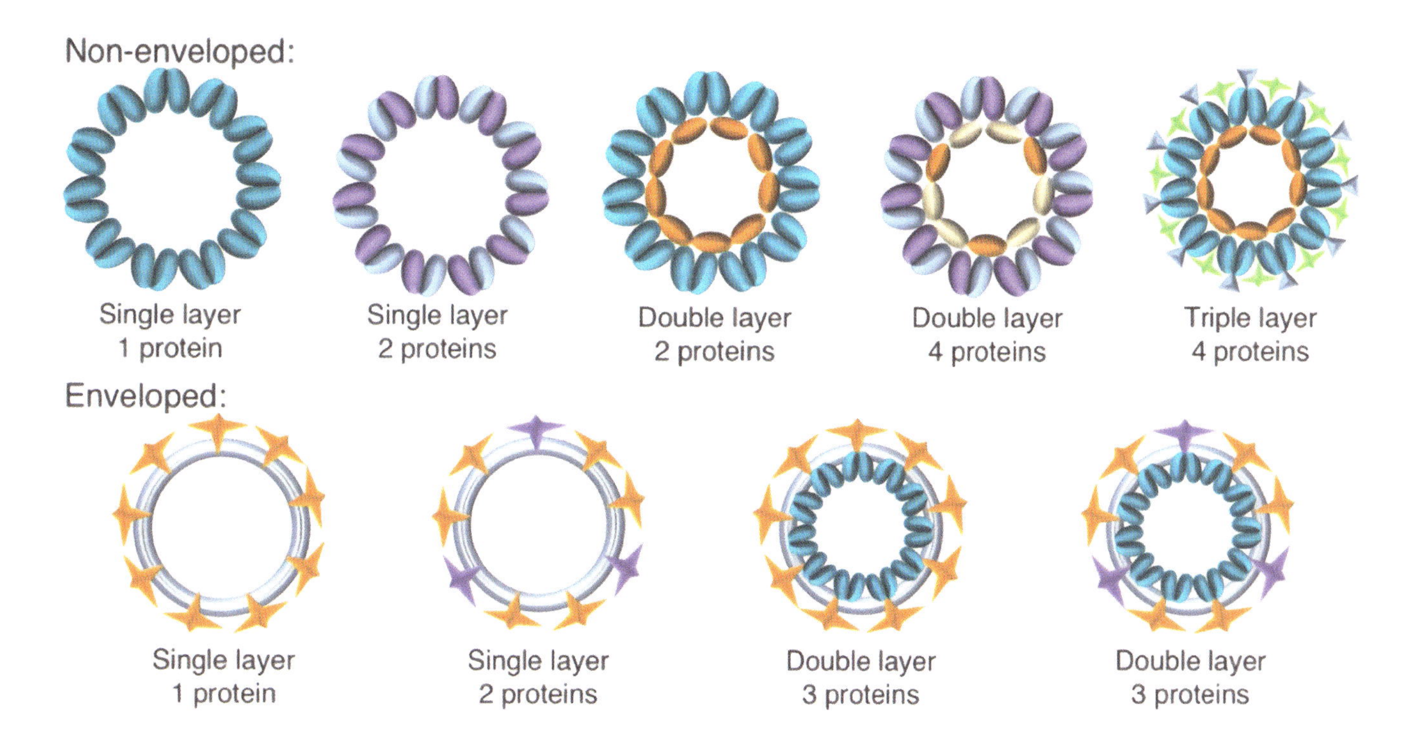

SARS-CoV-2 Virus Like Particles (VLPs)

Spike proteins are produced by our body following the injection of any vaccine designed for the virus, SARS-CoV-2. The sharp bumps or little crown that project outward, are the spike proteins. These allow the virus to enter the cells and begin a self-replication process. The result is Covid-19 disease, infection, and illness.

The proclamation by vaccine manufacturers is that once injected, spike proteins will stimulate our body to produce synthetic proteins of similar construct and behavior. Their claim is these synthetic proteins will resist the resultant infection, when in fact, the opposite is true. Spike proteins promote infection by facilitating the entry of the virus into the cell.

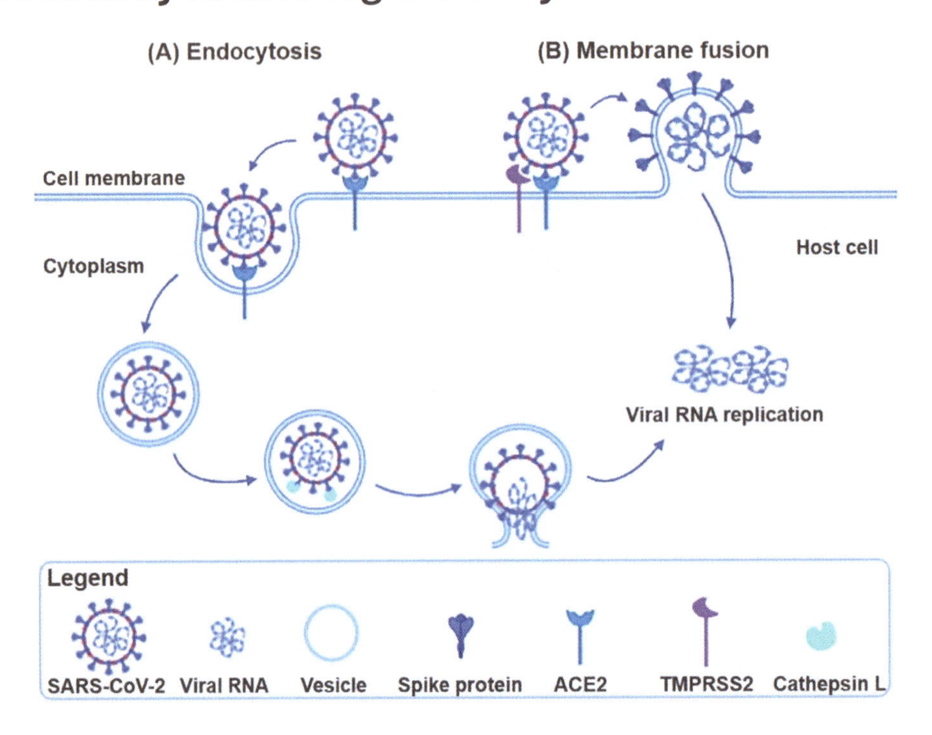

Naturally occuring ("wild") coronvirus spike proteins mimic proteins which regulate blood vessels and clotting. Spike proteins, both wild and synthetic, appear to to be shedding. They are being transmitted and assimilated through exhalation/inhalation, saliva, urine, seminal fluid, and feces.

What is Viral Shedding?

By Benedette Cuffari, M.Sc.

Reviewed by Sophia Coveney, B.Sc.

Excerpts:

Understanding the duration of viral shedding of severe acute respiratory syndrome coronavirus 2 (SARS-CoV-2), as well as how it relates to a positive or negative PCR test, is crucial to the implementation of effective public health efforts aimed towards controlling the spread of the virus.

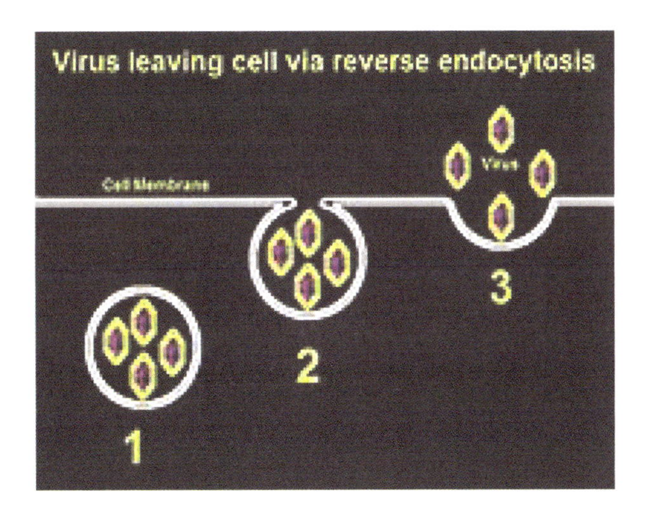

When an individual gets infected by a respiratory virus like SARS-CoV-2, the virus particles will bind to the various types of viral receptors, particularly the angiotensin-converting enzyme 2 (ACE2) receptors in the case of SARS-CoV-2, that line the respiratory tract. Throughout this ongoing process, infected individuals, who may not yet be experiencing any of the viral symptoms, are shedding viral particles while they talk, exhale, eat, and perform other normal daily activities.

Under normal circumstances, viral shedding will not persist for more than a few weeks; however, as researchers gain a more in-depth understanding of the viral clearance of SARS-CoV-2, they have found that certain populations will shed this virus for much longer durations. In fact, a growing amount of evidence indicates that the viral shedding of SARS-CoV-2 begins before a patient is symptomatic, peaks at the point of or shortly after symptom onset and can continue to be released even after the individual's symptoms have been resolved.

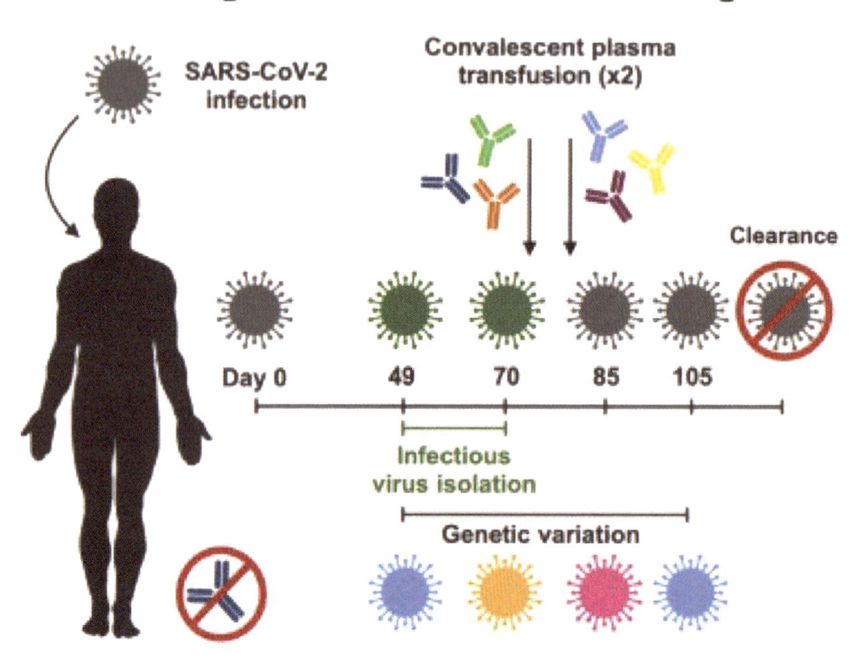

The duration of viral shedding can be used to categorize the infectivity of a person; therefore, this information is crucial in implementing effective infection prevention strategies, such as appropriate quarantine durations and mask requirements.

Currently, SARS-CoV-2 infection is confirmed with a positive polymerase chain reaction (PCR) test that can be conducted regardless of whether an individual is experiencing symptoms. Through such PCR tests, viral shedding of SARS-CoV-2 has been found to have a median duration of 12 to 20 days, with a persistence that can reach up to 63 days after initial symptom onset.

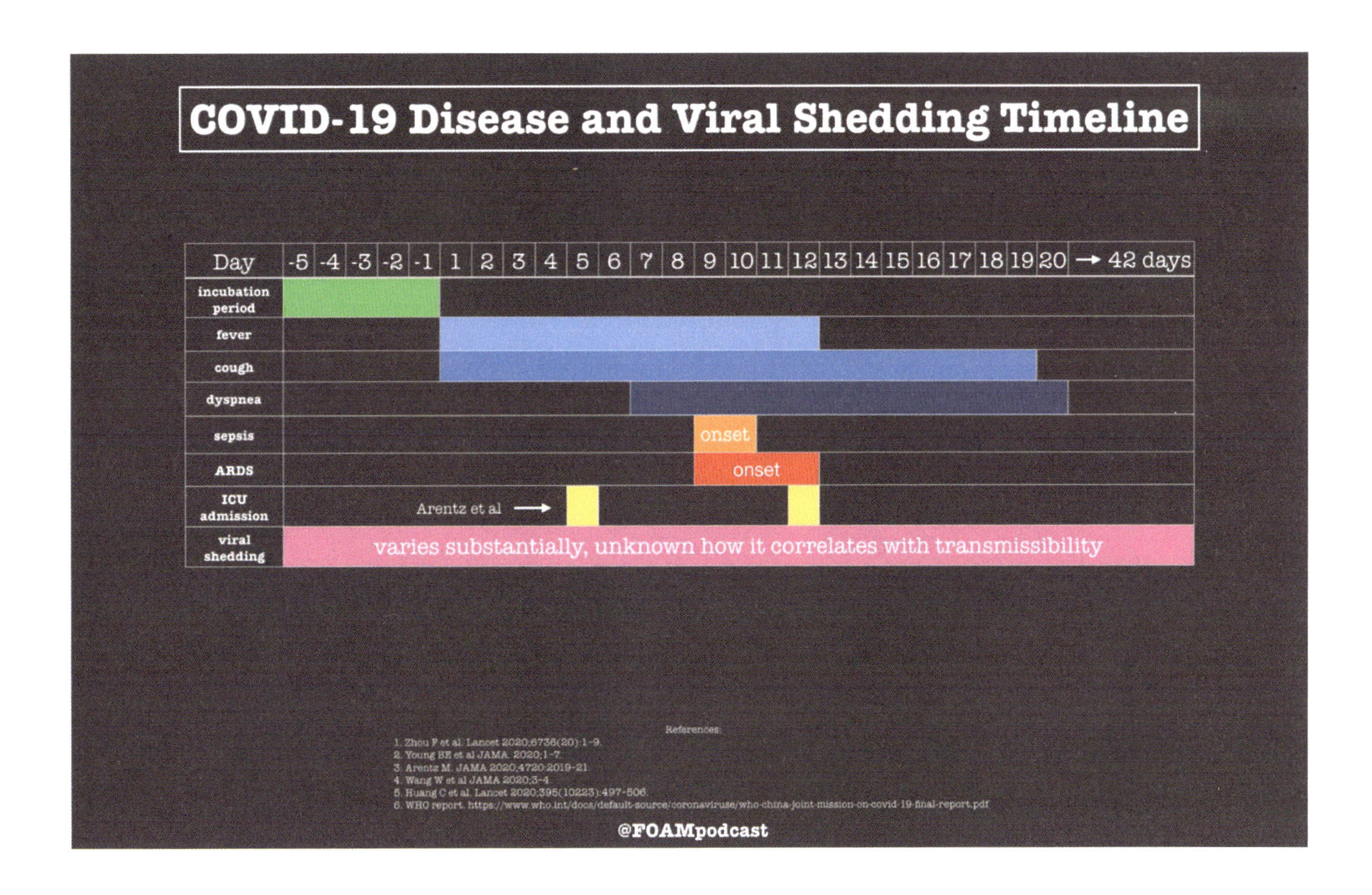

Whereas about 90% of mild cases have been found to clear the virus within an average of 10 days after symptom onset, individuals who have recovered from the severe disease have been found to have prolonged viral RNA shedding with a median duration of 31 days.

In addition to symptom severity being a predictive factor of viral shedding duration, the sampling location also appears to determine when peak viral loads occur. Within the upper respiratory tract (URT), for example, peak viral load appears to occur between days 4 and 6 following the onset of symptoms, within a week of symptom-onset, whereas peak viral loads within the lower respiratory tract appear to arise later.

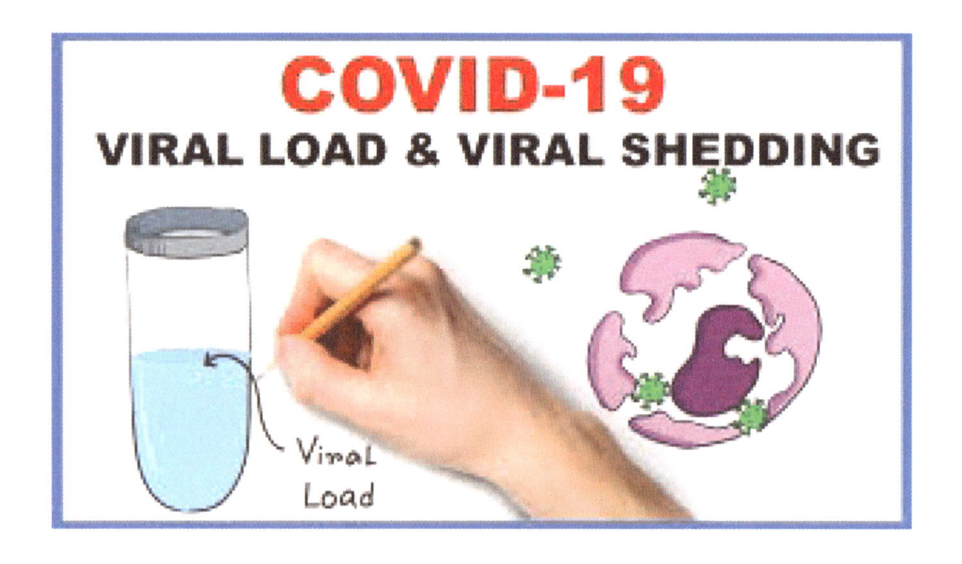

The viral shedding of SARS-CoV-2 also occurs within the gastrointestinal (GI) tract in the form of stool for up to 33 days after a negative PCR test; however, these viral loads appear to be less as compared to those identified within the respiratory tract and occur at a later time. Notably, the viral shedding of SARS-CoV-2 from the GI tract does not appear to have any correlation with disease severity.

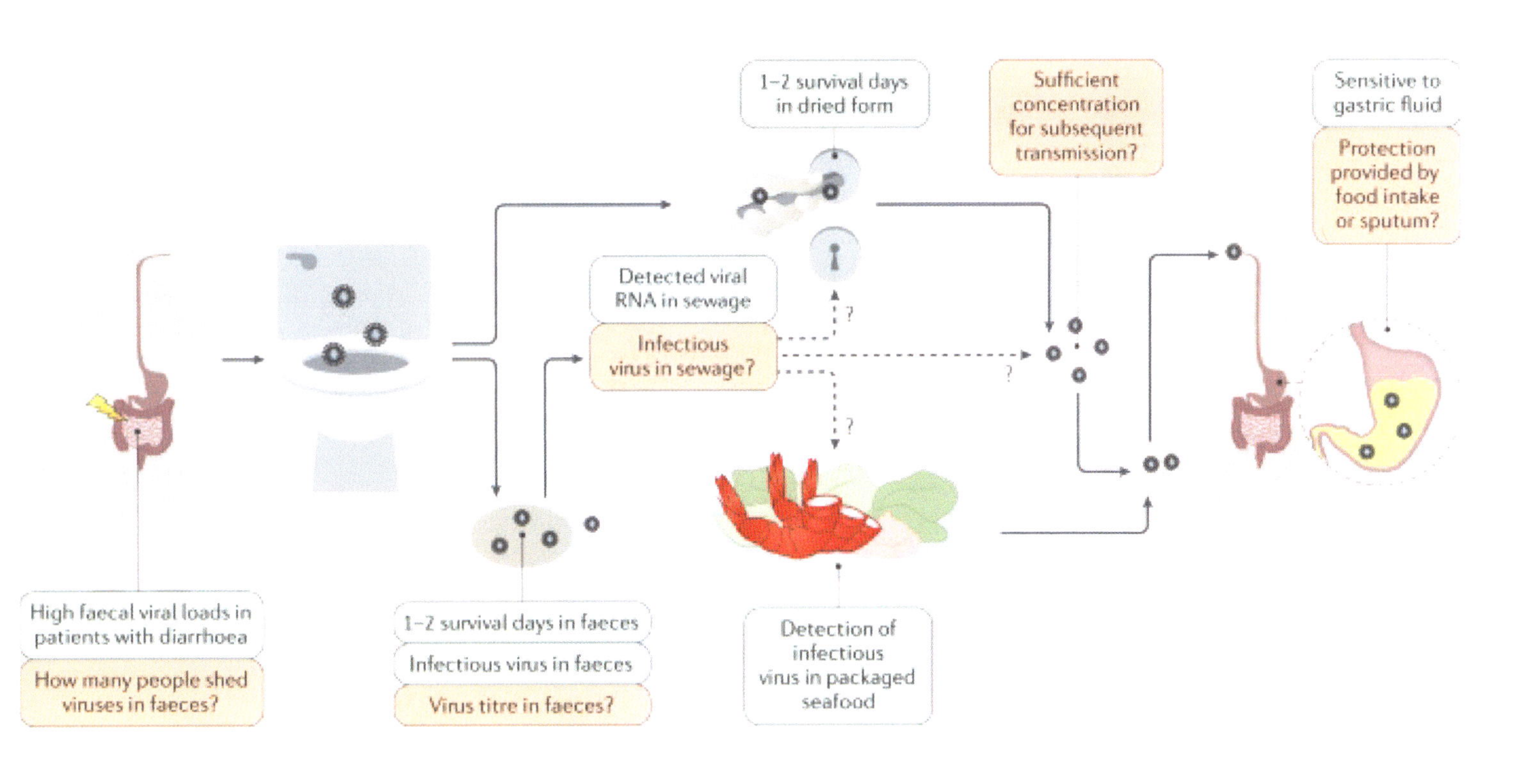

Unfortunately, there remains uncertain information on the proportion of SARS-CoV-2 cases that are asymptomatic. The reported values of asymptomatic cases can range anywhere from 1% to as high as 78%.

Moreover, it is unclear as to whether these "asymptomatic" cases are truly asymptomatic in the sense that these infected individuals will never experience any of the viral symptoms, or are rather presymptomatic, meaning that these individuals had no symptoms at the time of their positive PCR test but eventually developed symptoms later.

Even among presymptomatic patients, the higher level of SARS-CoV-2 viral shedding from the URT is a key factor in its high transmissibility, particularly when compared to its genetically similar predecessor SARS, which mainly occurred within the lower respiratory tract.

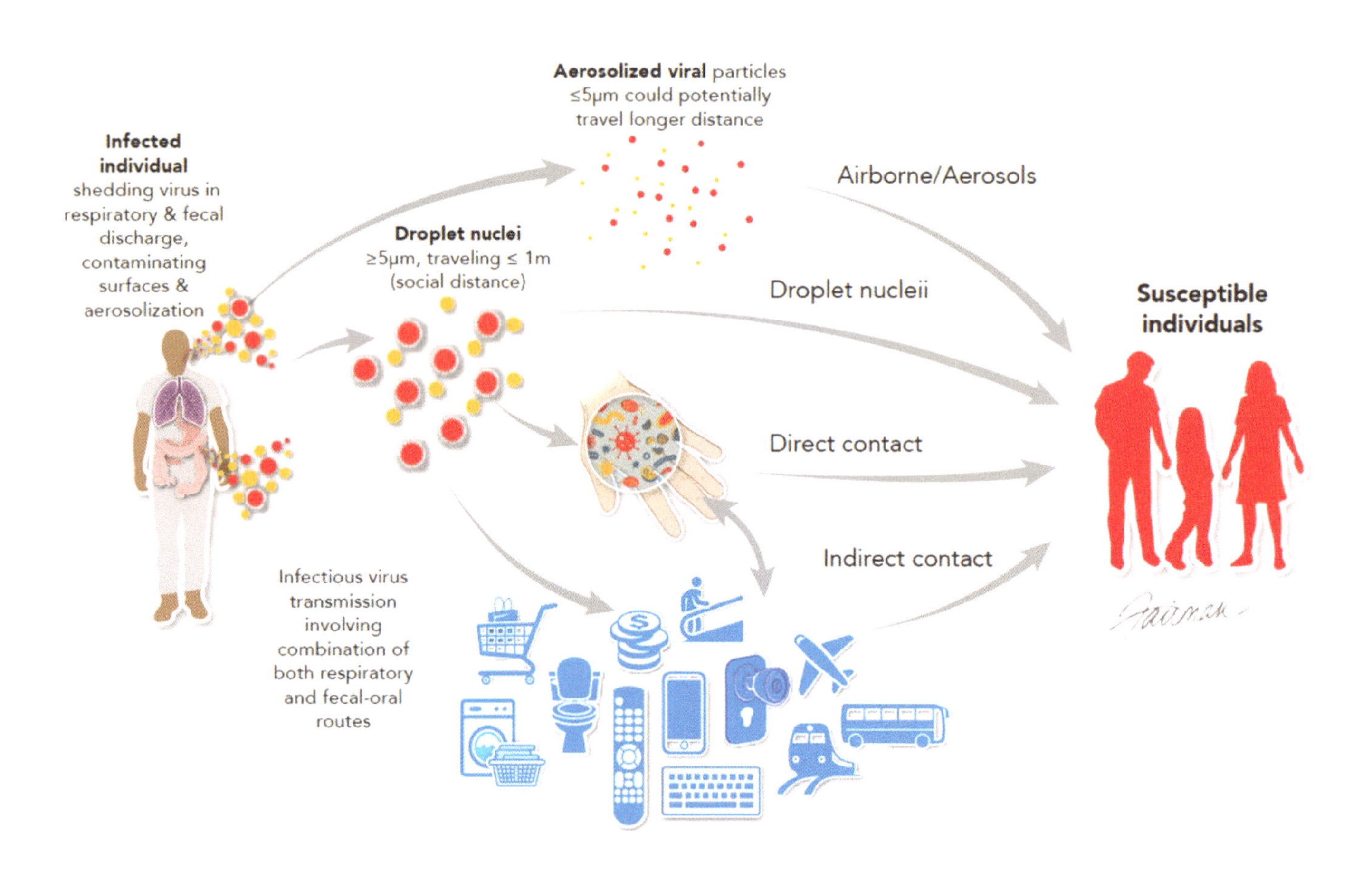

COVID19 – the spike protein and blood clotting

Dr. Malcolm Kendrick

3rd June 2021

Excerpts:

The spike protein

Then, of course, we have the spike protein to consider. If this is the thing that the immune system recognises and attacks – which it almost certainly is – then cells which are growing SARS-Cov2 inside them, which then express the spike protein on their surface as the virions escape, will be identified as 'the enemy'.

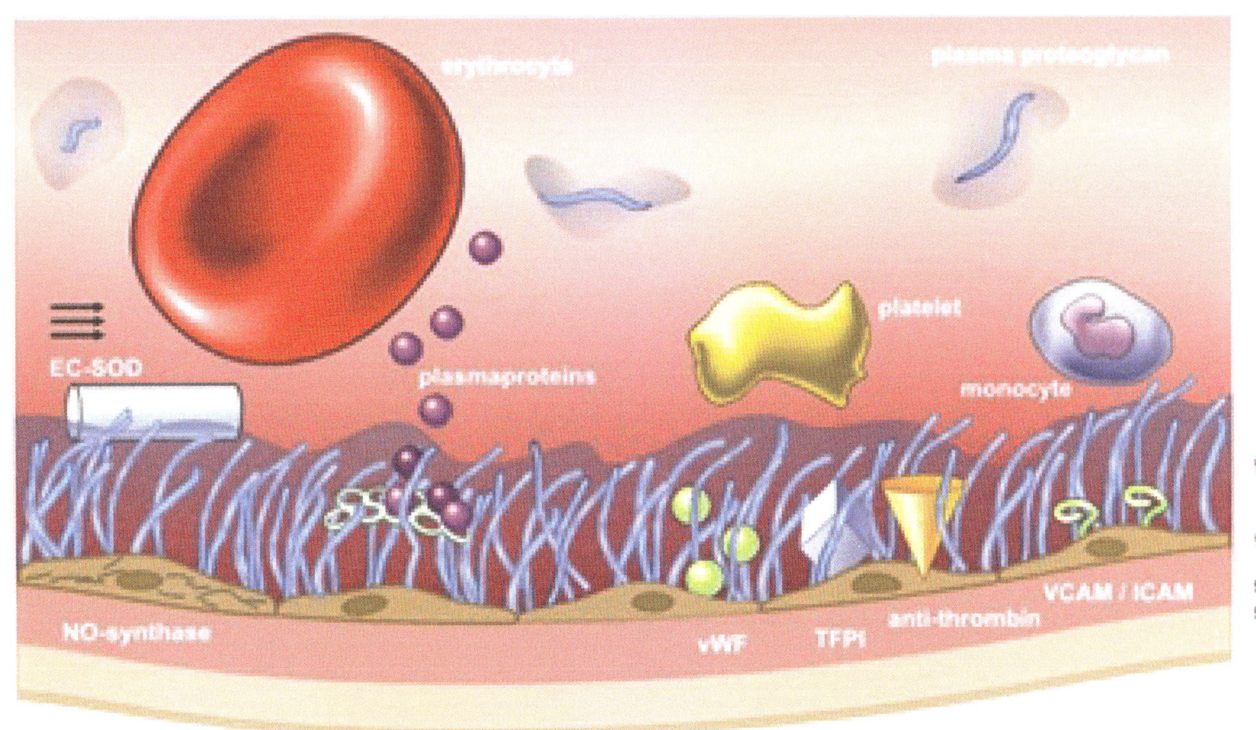

At which point, the immune system will start to attack the endothelium (and glycocalyx) in an attempt to wipe out the virus. This will tend to happen two or three weeks after the initial infection (sometimes sooner). This is after the immune system has had a real chance to identify the spike protein, then properly wind itself up to produce antibodies against it. This is the time of maximum attack on the endothelium.

This moment is often referred to as a cytokine storm. A point where every system in the immune system gets revved up and charges into action. At one point I wasn't sure if I really believed in the cytokine storm. But I do now think it is a real thing. It is almost certainly why steroids (which very powerfully reduce the immune response) have been found to reduce mortality in severely ill patients.

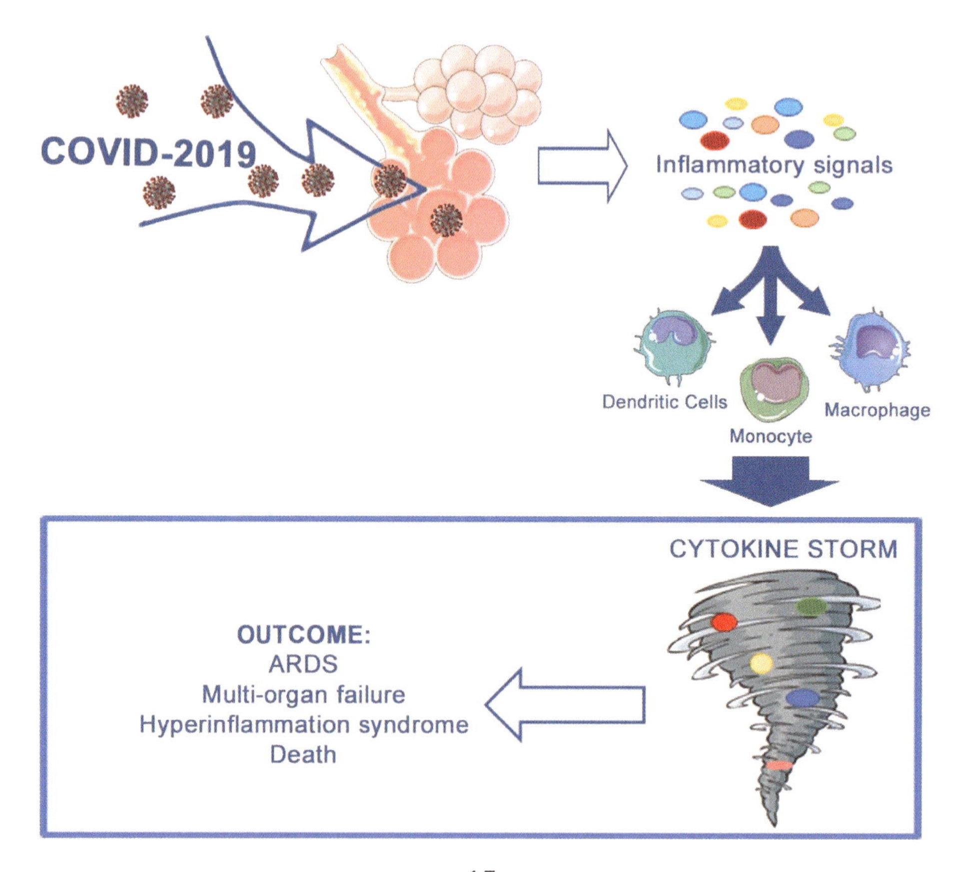

All of which means it may well be the body's own infectious disease defence system that creates much of the damage to the cardiovascular system. Not necessarily the virus itself. **(Emphasis added)**

Alternatively, it may be that the spike protein itself creates most of the blood clots. Here from the paper '*SARS-CoV-2 spike S1 subunit induces hypercoagulability.*'

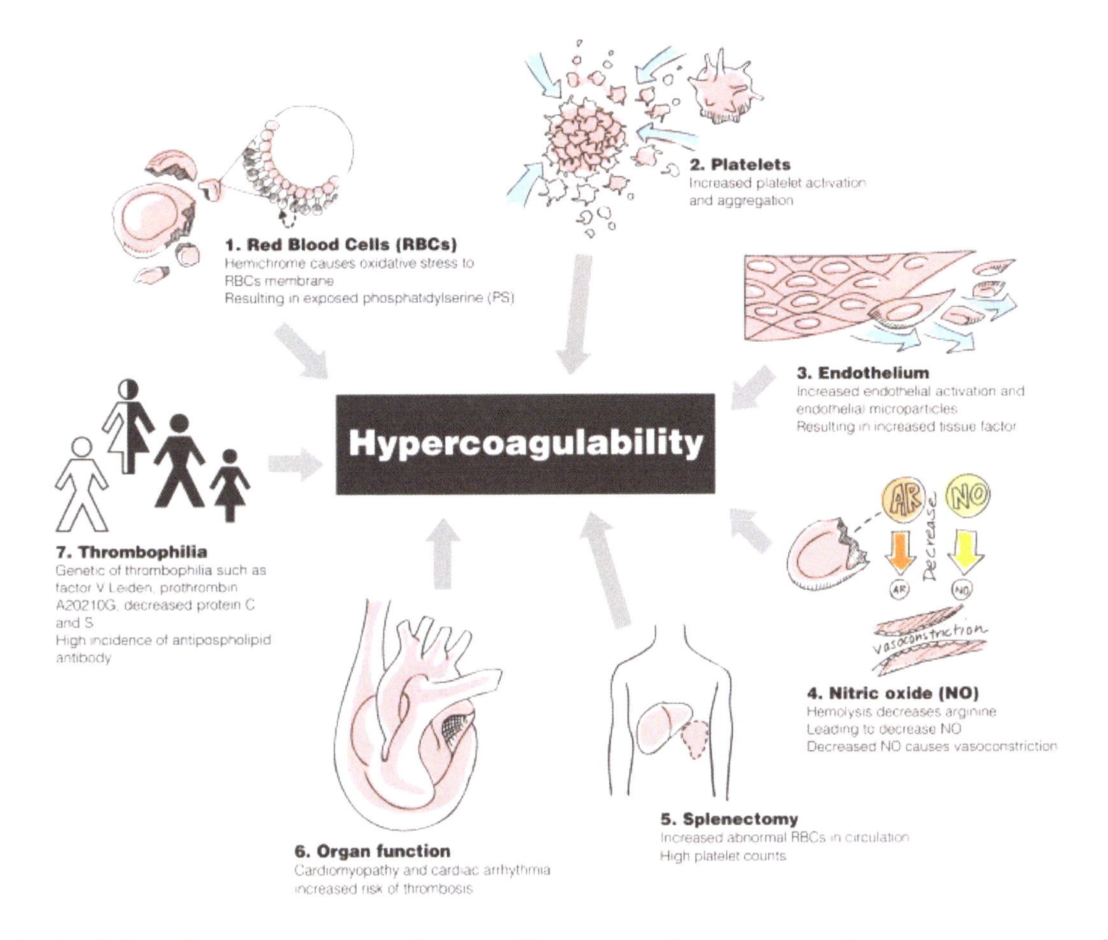

When whole blood was exposed to spike protein even at low concentrations, the erythrocytes*(red blood cells) *showed agglutination, hyperactivated platelets were seen, with membrane spreading and the formation of platelet-derived microparticles.

Translation. Introduce SARS-CoV2 spike proteins into bloodstream, and it makes it clot – fast. Which is a worry.

Vaccines

It is a worry because the entire purpose of vaccination against SARS-Cov2 is to force cells to manufacture the spike protein(s) and then send them out into the bloodstream.

We know that a very high percentage of the people who die following a COVID19 infection, die as result of blood clots. We also know that they can also suffer severe myocarditis (inflammation of the heart muscle), and suchlike.

We know that the spike protein can stimulate blood clots all by itself.

We know that the immune system attack on 'alien' proteins, such as the spike protein, can cause vasculitis.

We know that vaccines are designed to drive the rapid production of spike proteins that will enter the blood stream specifically to encounter immune cells, in order to create a powerful response that will lead to 'immunity' against future SARS-CoV2 infection.

We know that a number of people have died from blood clots following vaccination. To quote from the European Medicines Agency website report on the AZ COVID19 vaccine:

'The PRAC (pharmacovigilance risk assessment committee) noted that the blood clots occurred in veins in the brain (cerebral venous sinus thrombosis, CVST) and the abdomen (splanchnic vein thrombosis) and in arteries, together with low levels of blood platelets and sometimes bleeding.'

My concern at this point is that, yes, we have identified very rare manifestations of blood clotting: cerebral venous sinus thrombosis (CVST) and splanchnic (relating to the internal organs or viscera) vein thrombosis (SVT). These are so rare that it is unlikely that anything else – other than a novel vaccine – could have caused them. I have never seen a case and I had never even heard of them before COVID19 came along. And I have spent years studying the blood coagulation system, and vasculitis, and suchlike.

So, if someone is vaccinated, then has a cerebral venous sinus thrombosis, or a splanchnic vein thrombosis, this is almost certainly going to be noted and recorded – and associated with the vaccination. Fine.

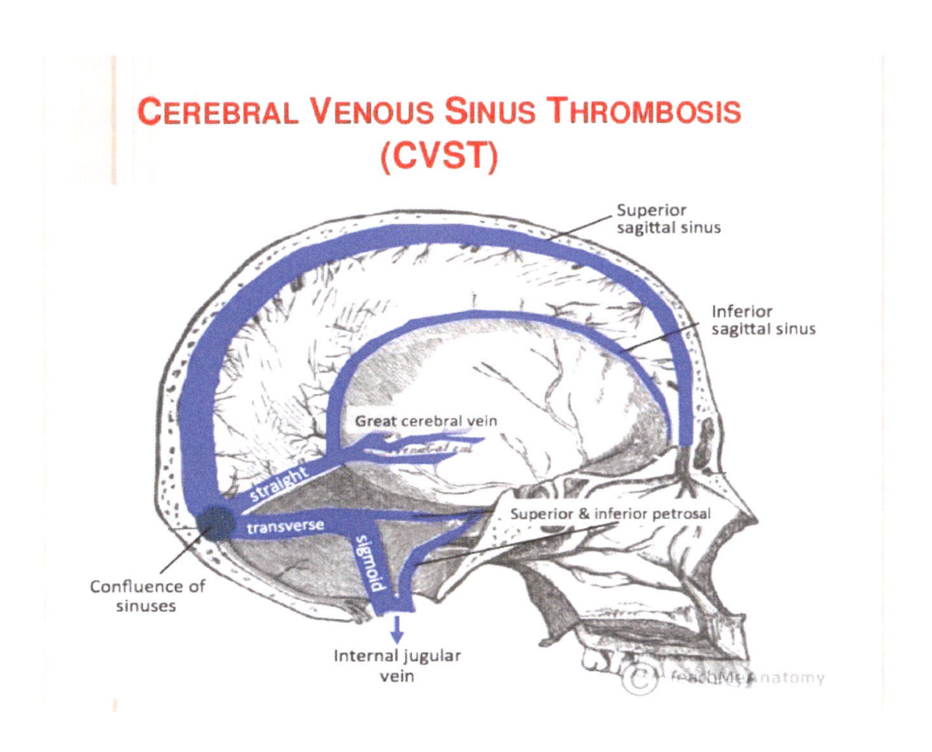

However, if there is an increase in vanishingly rare blood clots, could there also be an increase in other, far more common blood clots at the same time. If this was the case, then it would be far more difficult to spot this happening.

(Emphasis added)

Another major problem with vasculitis is that blood clots spring to life throughout the vascular system. This is because the blood is always ready to clot, at any time, and if you take away some of vital the anti-clotting mechanisms, the balance will be tilted firmly towards coagulation.

One of the most powerful anti-clotting mechanisms/systems is the protective layer that lines your entire vascular system, known as the glycocalyx. This is made up of glycoproteins (glucose and proteins stuck together).

In your blood vessels, the glycocalyx protrudes out from endothelial cells, the cells that line all your blood vessels, and into the bloodstream. The layer of glycocalyx contains many, many, anticoagulant factors. Below is a short list of all the things the glycocalyx does:

- Forms the interface between the vessel wall and moving blood.
- Acts as the exclusion zone between blood cells and the endothelium.
- Acts as a barrier against leakage of fluid, proteins**and lipids**across the vascular wall.
- Interacts dynamically with blood constituents.
- Acts as the "molecular sieve" for plasma proteins.
- Modulates adhesion of inflammatory cells and platelets to the endothelial surface.
- Functions as a sensor and mechano-transducer of the fluid shear forces to which the endothelium is exposed; thus, the glycocalyx mediates shear-stress-dependent nitric oxide production.
- Retains protective enzymes (e.g., superoxide dismutase).
- Retains anticoagulation factors, e.g.: Tissue factor inhibitor, Protein C, Nitric Oxide (NO), Antithrombin.

Anyway, if you damage the glycocalyx, or damage the underlying endothelial cells that synthesizes the glycocalyx layer, you will tip the balance very strongly towards the creation of blood clots. These can then then stick to the artery, or vein, wall. Sometimes they will fully block a blood vessel, leading to such things as a stroke or heart attack. **(Emphasis added)**

The interaction between vasculitis and thrombosis has been a relatively unexplored area of medicine. But it remains critically important in many diseases:

'The relationship between inflammation and thrombosis is not a recent concept, but it has been largely investigated only in recent years. Nowadays inflammation-induced thrombosis is considered to be a feature of systemic autoimmune diseases such as Systemic Lupus Erythematosus (SLE), Rheumatoid Arthritis (RA), or Sjogren's Syndrome (SS)2.

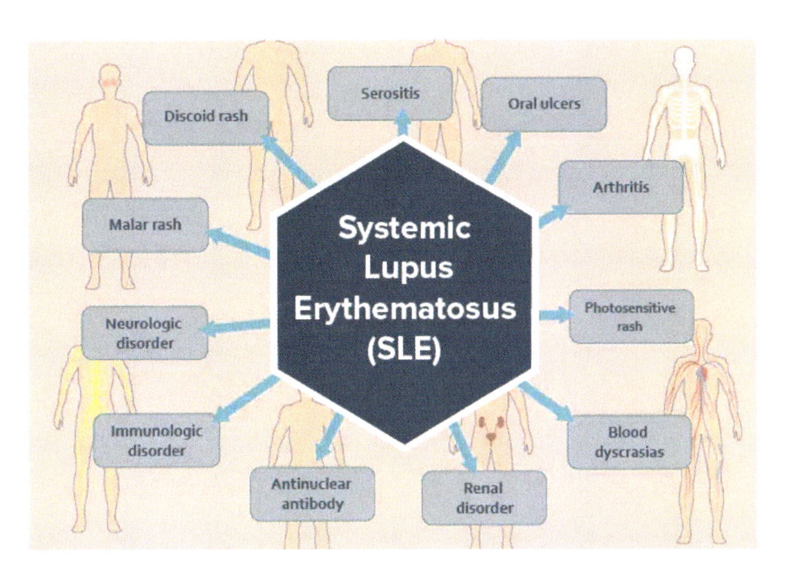

In super-short version. If you damage the lining of blood vessel walls, blood clots are far more likely to form. Very often, the damage is caused by the immune system going on the attack, damaging blood vessel walls, and removing several of the anti-clotting mechanisms. **(Emphasis added)**

Worse Than the Disease? Reviewing Some Possible Unintended Consequences of the mRNA Vaccines Against COVID-19

Stephanie Seneff and Greg Nigh Computer Science and Artificial Intelligence Laboratory, MIT, Cambridge MA, 02139,

Naturopathic Oncology, Immersion Health, Portland, OR 97214, USA

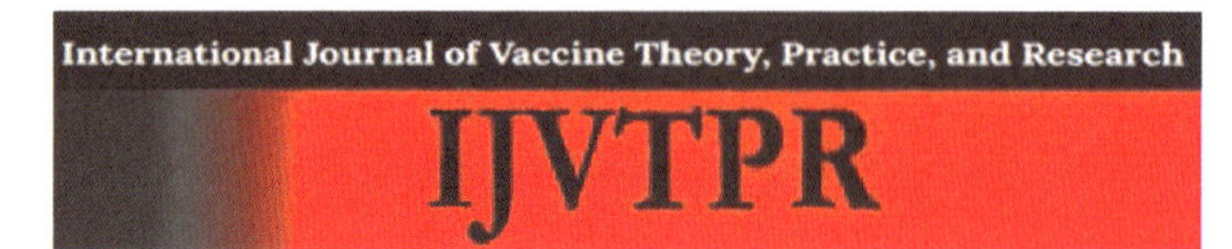

Excerpts:

It has been claimed that mRNA-based vaccines are safer than DNA-vectored vaccines that work by incorporating the genetic code for the target antigenic protein into a DNA virus, because the RNA cannot become inadvertently incorporated into the human genome. However, it is not at all clear that this is true. The classic model of DNA → RNA → protein is now known to be false. It is now indisputable that there is a large class of viruses called retroviruses (e.g. HIV-1) that carry genes that reverse transcribe RNA back into complementary DNA (cDNA). In 1975, Howard Temin, Renato Dulbecco, and David Baltimore shared the Nobel Prize in Physiology or Medicine in 1975 for their discovery of reverse transcriptase and its synthesis by retroviruses (such as human immunodeficiency virus (HIV)) to derive DNA from RNA (Temin and Mizutani, 1970, Baltimore, 1970).

Much later, it was discovered that reverse transcriptase is not unique to retroviruses. More than a third of the human genome is devoted to mysterious mobile DNA elements called SINEs and LINEs (short and long interspersed nuclear elements, respectively). LINEs provide reverse transcriptase capabilities to convert RNA into DNA, and SINEs provide support for integrating the DNA into the genome. Thus, these elements provide the tools needed to convert RNA into DNA and incorporate it into the genome so as to maintain the new gene through future generations (Weiner, 2002).

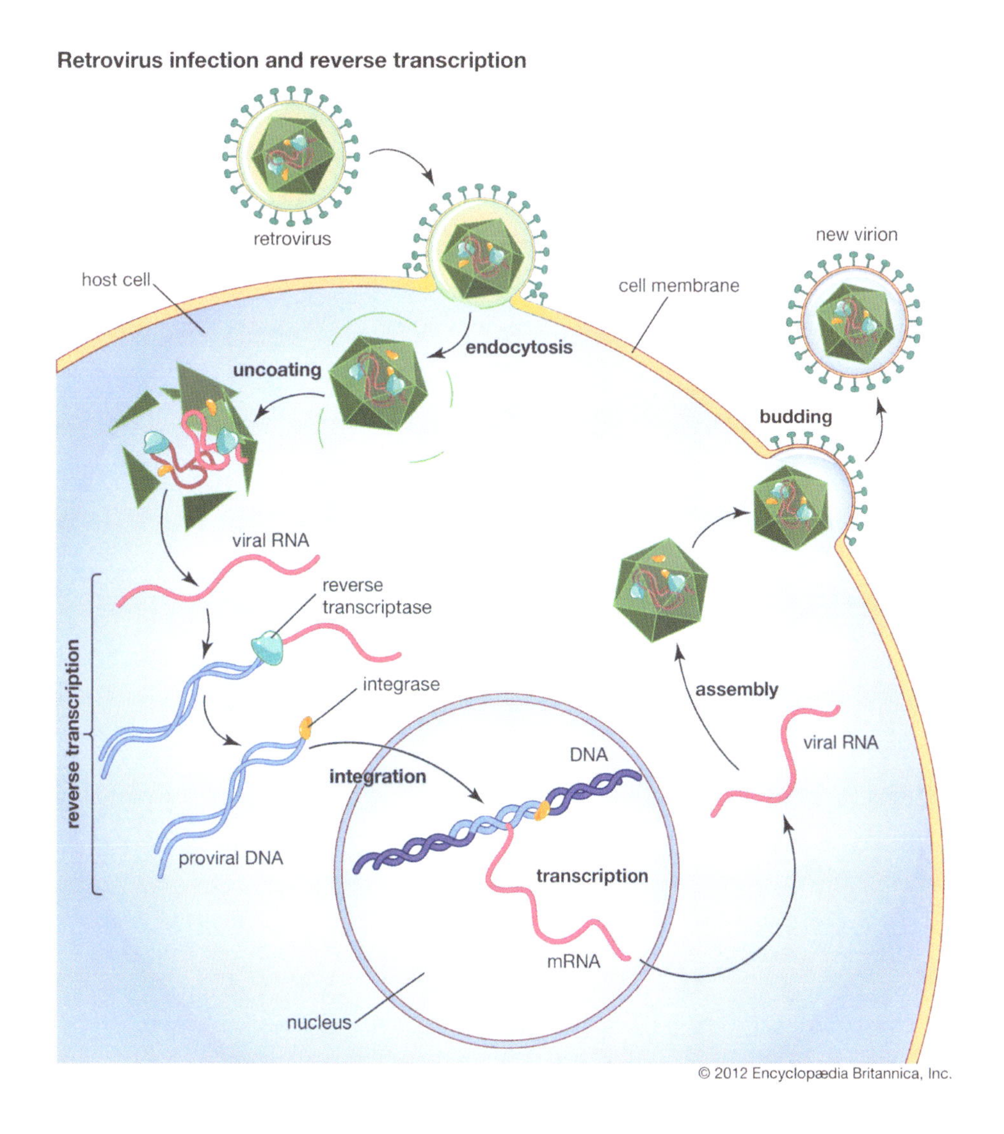

SINEs and LINEs are members of a larger class of genetic elements called retrotransposons. Retrotransposons can copy and paste their DNA to a new site in the genome via an RNA intermediate, while possibly introducing genetic alterations in the process (Pray, 2008). Retrotransposons, also known as "jumping genes," were first identified by the geneticist Barbara McClintock of Cold Spring Harbor Laboratory in New York, over 50 years ago (McClintock, 1965). Much later, in 1983, she was recognized with a Nobel prize for this work.

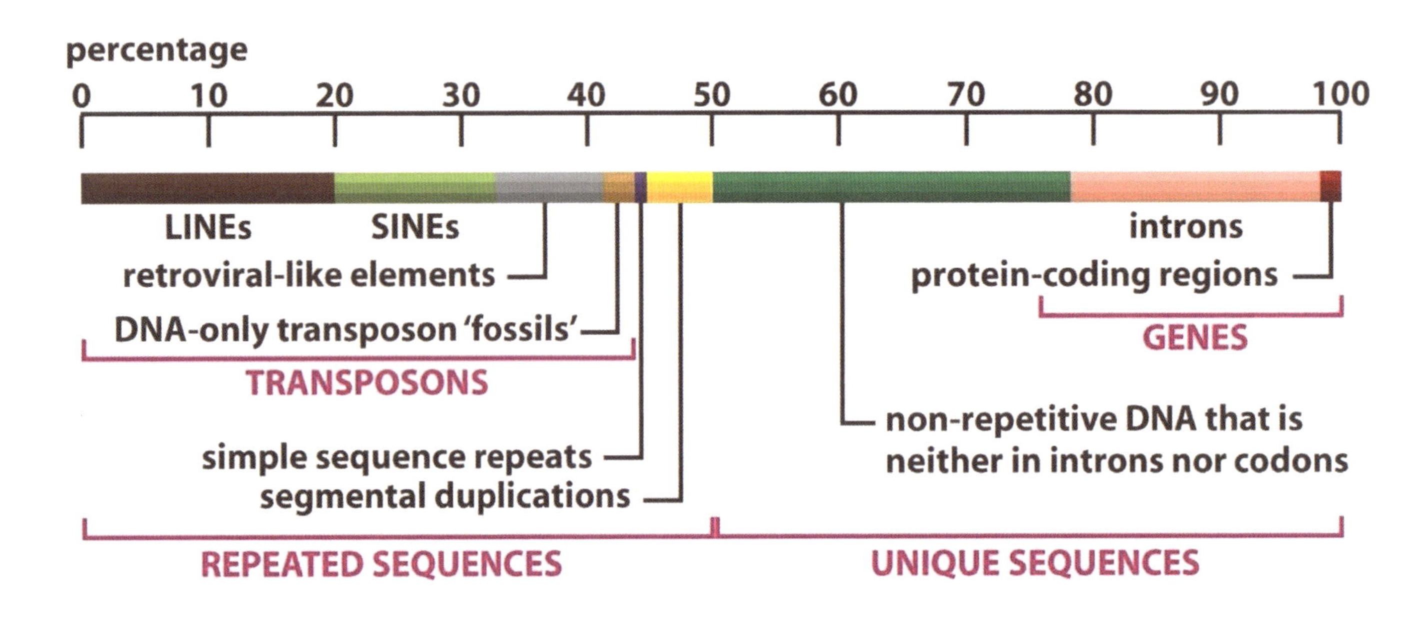

Remarkably, retrotransposons seem to be able to expand their domain from generation to generation. LINEs and SINEs collaborate to invade new genomic sites through translation of their DNA to RNA and back to a fresh copy of DNA, which is then inserted at an AT-rich region of the genome.

These LINEs and SINEs had long been considered to be "junk" DNA, an absurd idea that has now been dispelled, as awareness of their critical functions has grown. In particular, it has now become clear that they can also import RNA from an exogenous source into a mammalian host's DNA. Retroviral-like repeat elements found in the mouse genome called intracisternal A particles (IAPs) have been shown to be capable of incorporating viral RNA into the mouse genome. Recombination between an exogenous nonretroviral RNA virus and an IAP retrotansposon resulted in reverse transcription of the viral RNA and integration into the host's genome (Geuking et al., 2009).

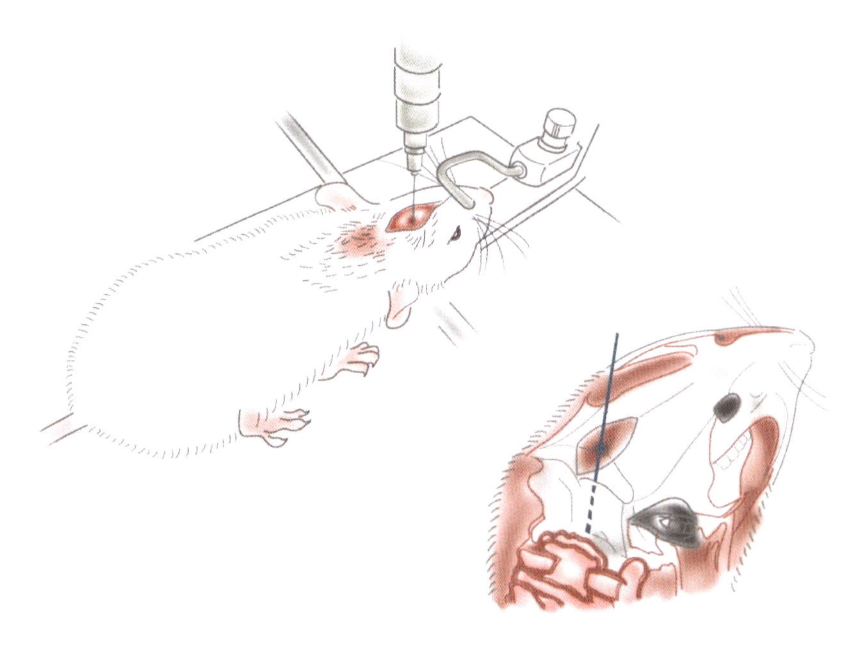

Furthermore, as we shall see later, the mRNA in the new SARS-CoV-2 vaccines could also get passed on from generation to generation, with the help of LINEs expressed in sperm, via nonintegrated cDNA encapsulated in plasmids. The implications of this predictable phenomenon are unclear, but potentially far-reaching.

There is also a concern that the RNA in the mRNA vaccines could be transferred into the human genome with assistance from retroviruses. Retroviruses are a class of viruses that maintain their genomic information in the form of RNA, but that possess the enzymes needed to reverse transcribe their RNA into DNA and insert it into a host genome. They then rely on existing natural tools from the host to produce copies of the virus through translation of DNA back into RNA and to produce the proteins that the viral RNA codes for and assemble them into a fresh viral particle (Lesbats et al., 2016).

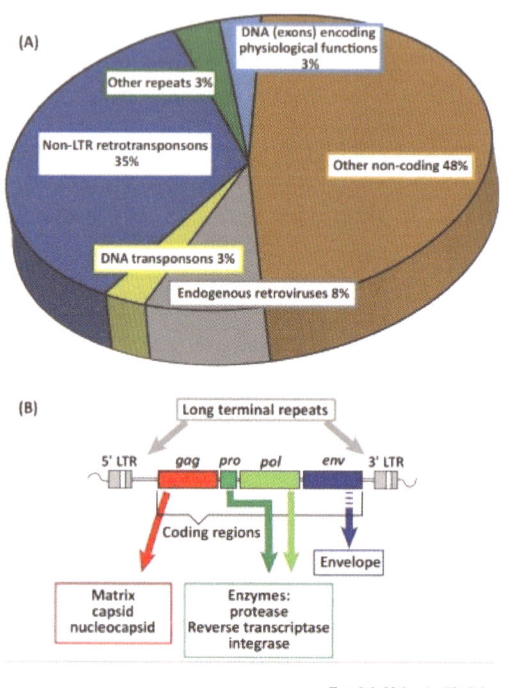

Trends in Molecular Medicine

Human endogenous retroviruses (HERVs) are benign sections in the DNA of humans that closely resemble retroviruses, and that are believed to have become permanent sequences in the human genome through a process of integration from what was originally an exogenous retrovirus. Endogenous retroviruses are abundant in all jawed vertebrates and are estimated to occupy 5-8% of the human genome.

The protein syncytin, which has become essential for placental fusion with the uterine wall and for the fusion step between the sperm and the egg at fertilization, is a good example of an endogenous retroviral protein. Syncytin is the envelope gene of a recently identified human endogenous defective retrovirus, HERV-W (Mi et al., 2000).

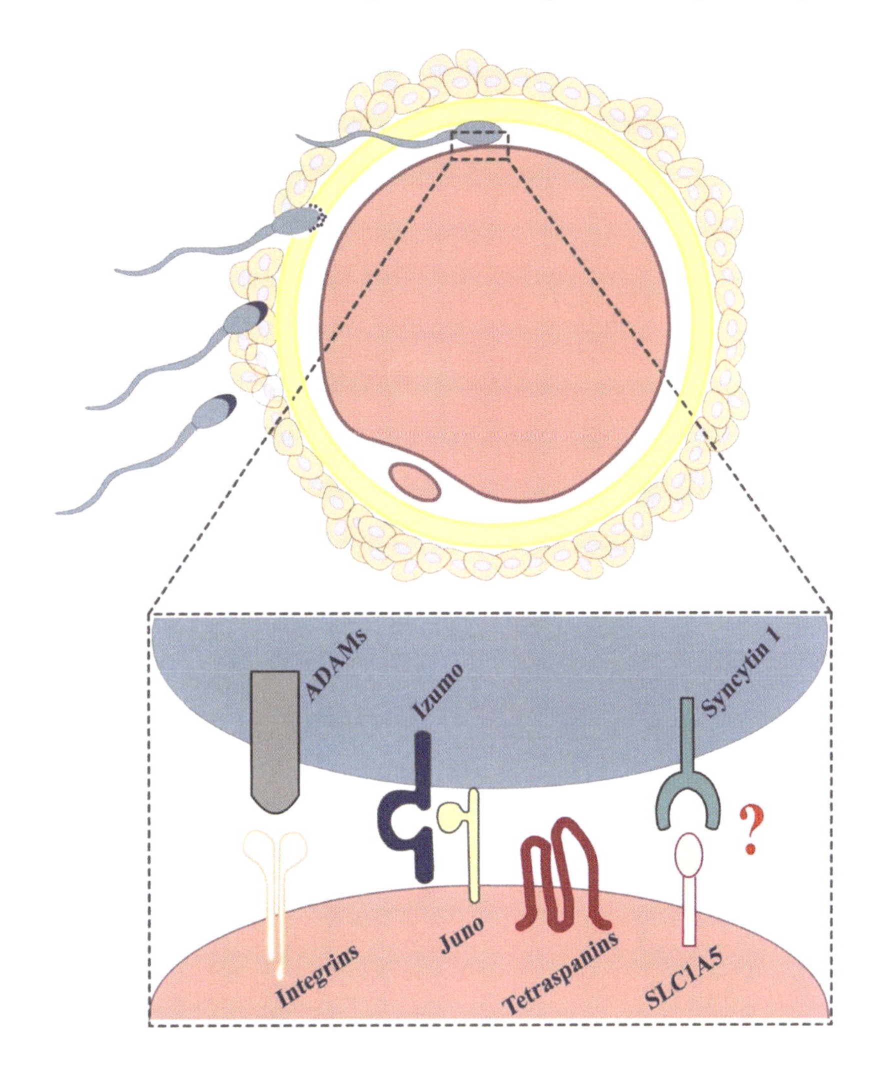

During gestation, the fetus expresses high levels of another endogenous retrovirus, HERV-R, and it appears to protect the fetus from immune attack from the mother (Luganini and Gribaudo, 2020). Endogenous retroviral elements closely resemble retrotransposons. Their reverse transcriptase, when expressed, has the theoretical capability to convert spike protein RNA from the mRNA vaccines into DNA.

Humans are colonized by a large collection of exogenous retroviruses that in many cases cause no harm to the host, and may even be symbiotic (Luganini and Gribaudo, 2020). Exogenous viruses can be converted to endogenous viruses (permanently incorporated into host DNA) in the laboratory, as demonstrated by Rudolf Jaenisch (Jaenisch, 1976), who infected preimplantation mouse embryos with the Moloney murine leukemia virus (M-MuLV).

The mice generated from these infected embryos developed leukemia, and the viral DNA was integrated into their germ line and transmitted to their offspring. Besides the incorporation of viral DNA into the host genome, it was also shown as early as 1980 that DNA plasmids could be microinjected into the nuclei of mouse embryos to produce transgenic mice that breed true (Gordon et al., 1980).

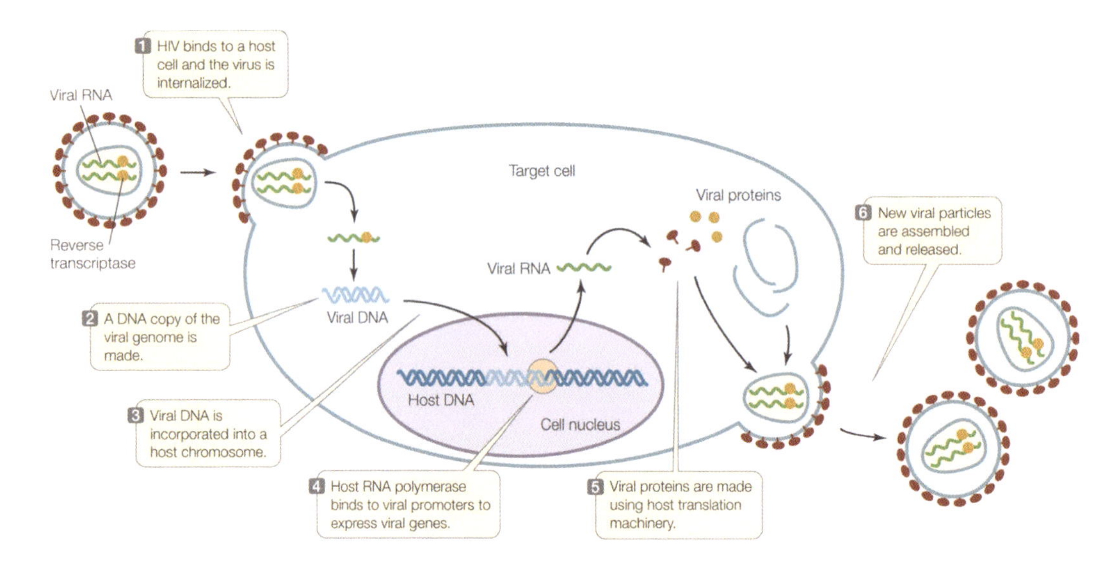

The plasmid DNA was incorporated into the nuclear genome of the mice through existing natural processes, thus preserving the newly acquired genetic information in the offspring's genome. This discovery has been the basis for many genetic engineering experiments on transgenic mice engineered to express newly acquired human genes since then (Bouabe and Okkenhaug, 2013).

Unprecedented

Many aspects of Covid-19 and subsequent vaccine development are unprecedented for a vaccine deployed for use in the general population. Some of these include the following.

1. First to use PEG (polyethylene glycol) in an injection (see text)

2. First to use mRNA vaccine technology against an infectious agent

3. First time Moderna has brought any product to market

4. First to have public health officials telling those receiving the vaccination to expect an adverse reaction

5. First to be implemented publicly with nothing more than preliminary efficacy data (see text)

6. First vaccine to make no clear claims about reducing infections, transmissibility, or deaths

7. First coronavirus vaccine ever attempted in humans

8. First injection of genetically modified polynucleotides in the general population

The mRNA vaccines are the culmination of years of research in exploring the possibility of using RNA encapsulated in a lipid particle as a messenger. The host cell's existing biological machinery is co-opted to facilitate the natural production of protein from the mRNA. The field has blossomed in part because of the ease with which specific oligonucleotide DNA sequences can be synthesized in the laboratory without the direct involvement of living organisms. This technology has become commoditized and can be done at large-scale, with relatively low cost. Enzymatic conversion of DNA to RNA is also straightforward, and it is feasible to isolate essentially pure single-strand RNA from the reaction soup (Kosuri and Church, 2014).

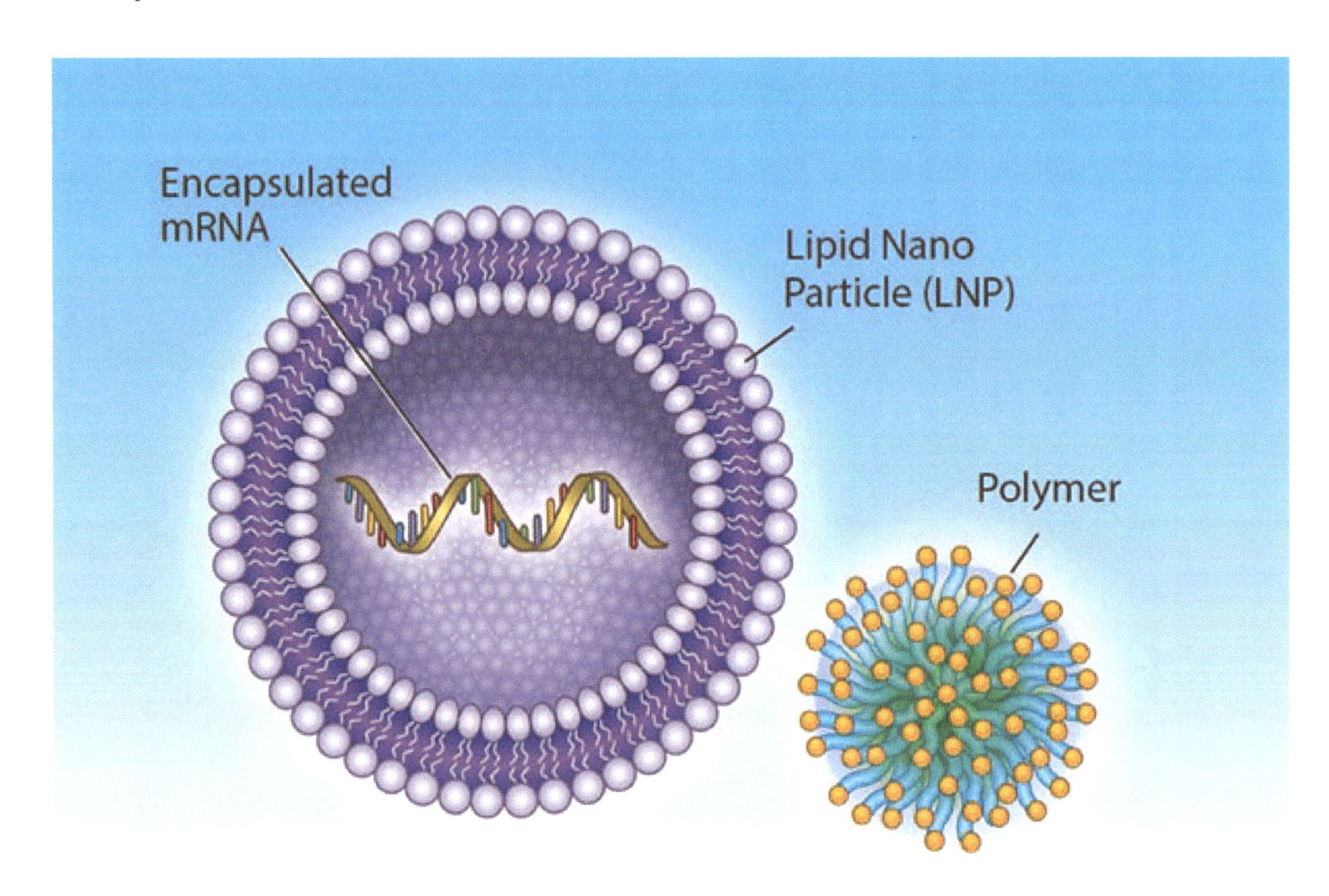

In the early phase of nucleotide-based gene therapy development, there was considerably more effort invested in gene delivery through DNA plasmids rather than through mRNA technology. Two major obstacles for mRNA are its transient nature due to its susceptibility to breakdown by RNAses, as well as its known power to invoke a strong immune response, which interferes with its transcription into protein. Plasmid DNA has been shown to persist in muscle up to six months, whereas mRNA almost certainly disappears much sooner. For vaccine applications, it was originally thought that the immunogenic nature of RNA could work to an advantage, as the mRNA could double as an adjuvant for the vaccine, eliminating the arguments in favor of a toxic additive like aluminum. However, the immune response results not only in an inflammatory response but also the rapid clearance of the RNA and suppression of transcription. So this idea turned out not to be practical.

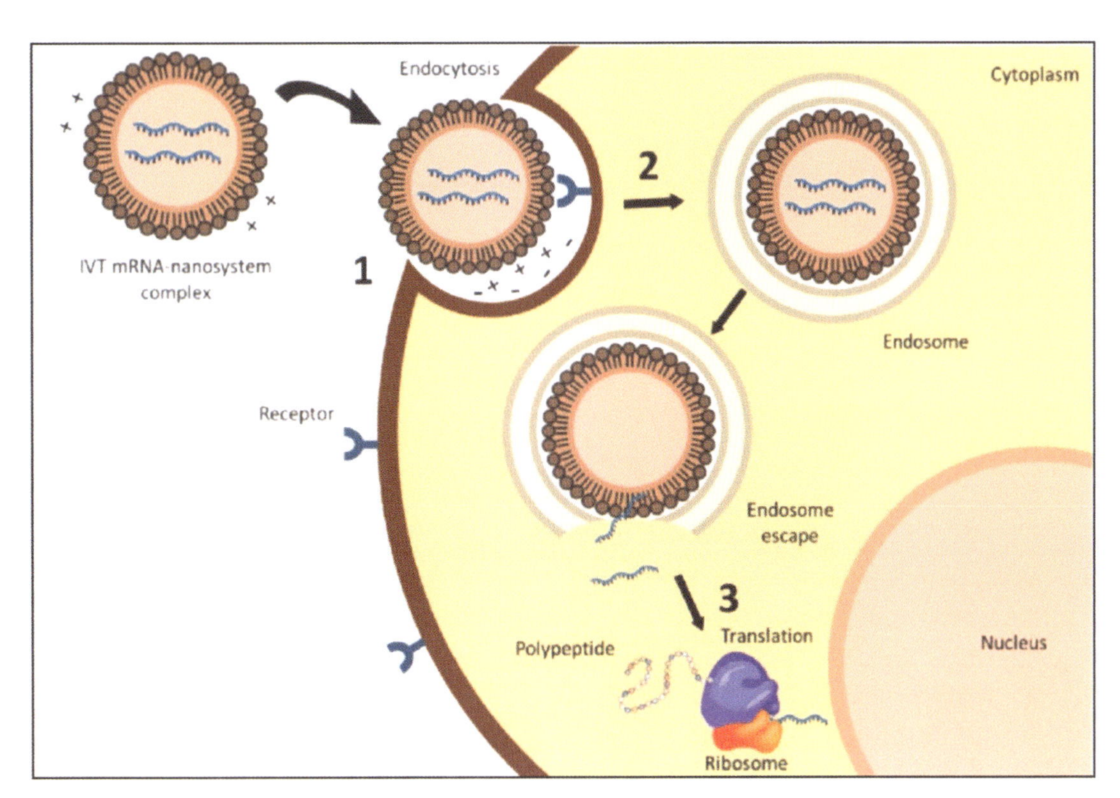

There was an extensive period of time over which various ideas were explored to try to keep the mRNA from breaking down before it could produce protein. A major advance was the realization that substituting methyl-pseudouridine for all the uridine nucleotides would stabilize RNA against degradation, allowing it to survive long enough to produce adequate amounts of protein antigen needed for immunogenesis (Liu, 2019).

This form of mRNA delivered in the vaccine is never seen in nature, and therefore has the potential for unknown consequences. **(Emphasis added)**

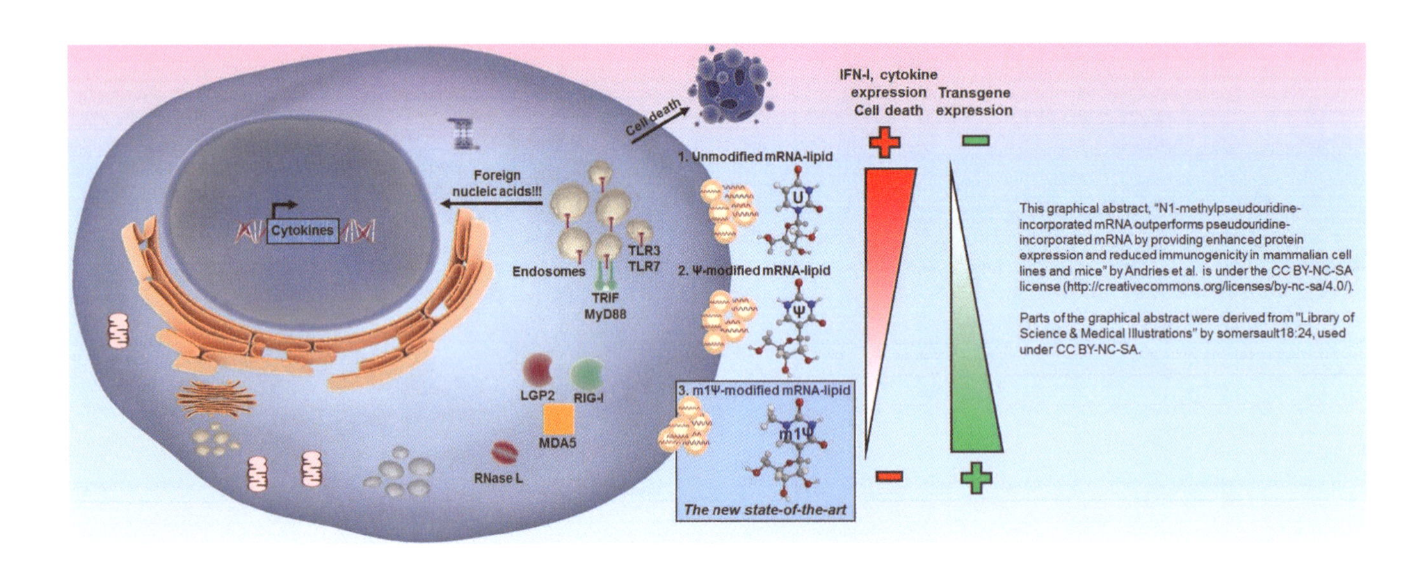

This graphical abstract, "N1-methylpseudouridine-incorporated mRNA outperforms pseudouridine-incorporated mRNA by providing enhanced protein expression and reduced immunogenicity in mammalian cell lines and mice" by Andries et al. is under the CC BY-NC-SA license (http://creativecommons.org/licenses/by-nc-sa/4.0/).

Parts of the graphical abstract were derived from "Library of Science & Medical Illustrations" by somersault18:24, used under CC BY-NC-SA.

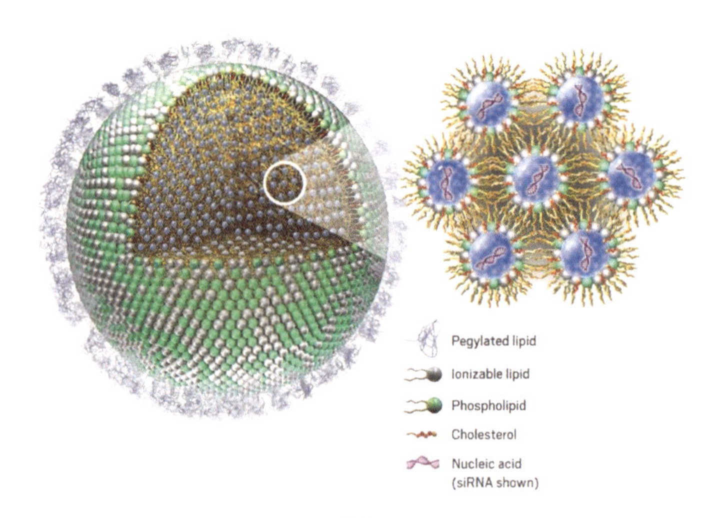

The Pfizer-BioNTech and Moderna mRNA vaccines are based on very similar technologies, where a lipid nanoparticle encloses an RNA sequence coding for the full-length SARS-CoV-2 spike protein. In the manufacturing process, the first step is to assemble a DNA molecule encoding the spike protein. This process has now been commoditized, so it's relatively straightforward to obtain a DNA molecule from a specification of the sequence of nucleotides (Corbett et al., 2020). Following a cell-free in vitro transcription from DNA, utilizing an enzymatic reaction catalyzed by RNA polymerase, the single-stranded RNA is stabilized through specific nucleoside modifications, and highly purified.

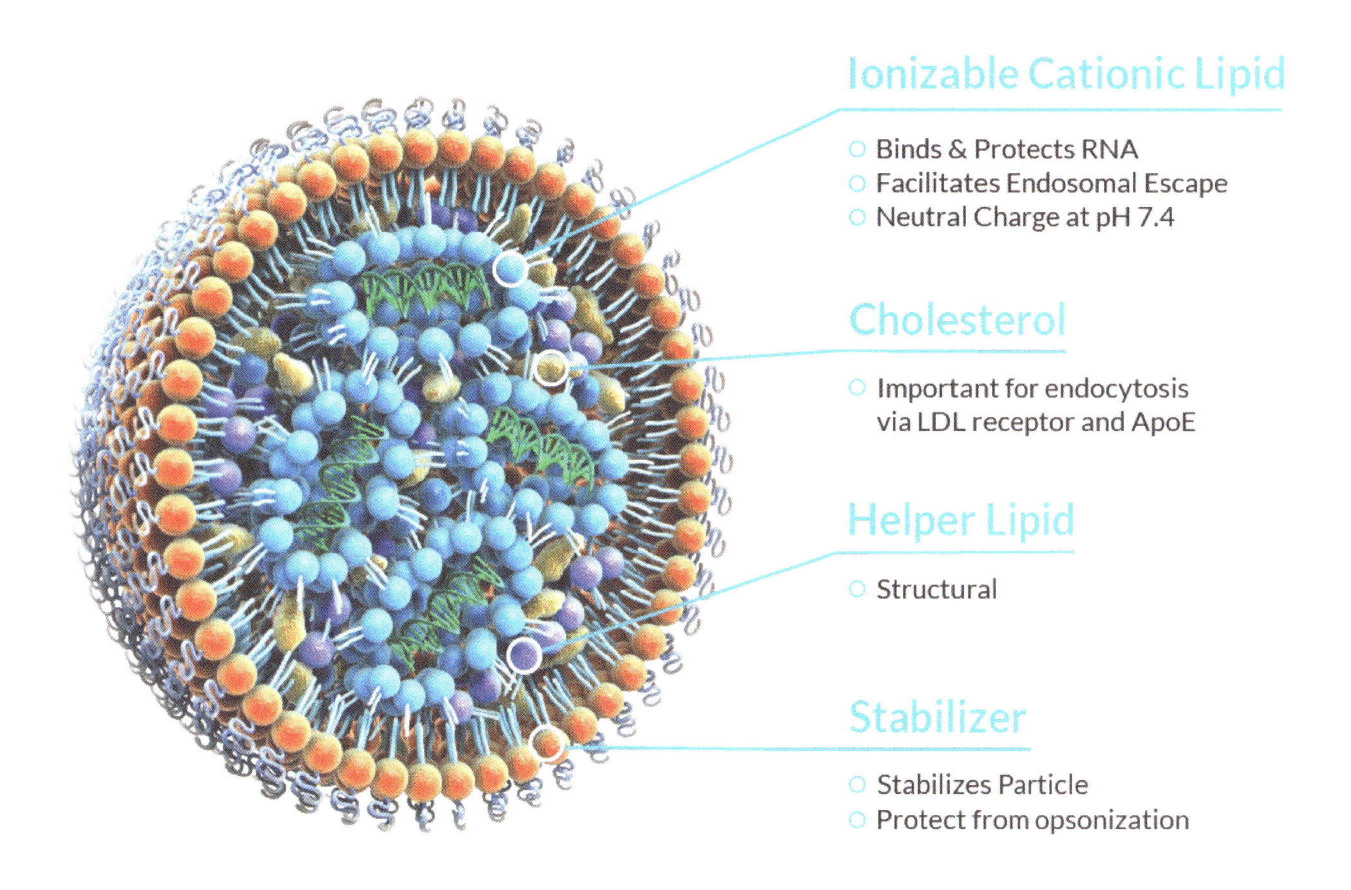

Researchers in China published a report in Nature in August 2020 in which they presented data on several experimental mRNA vaccines where the mRNA coded for various fragments and proteins in the SARS-CoV-2 virus. They tested three distinct vaccine formulations for their ability to induce an appropriate immune response in mice. The three structural proteins, S (spike), M and E are minimal requirements to assemble a "virus-like particle" (VLP). Their hypothesis was that providing M and E as well as the S spike protein in the mRNA code would permit the assembly of VLPs that might elicit an improved immune response, because they more closely resemble the natural virus than S protein exposed on the surface of cells that have taken up only the S protein mRNA from the vaccine nanoparticles. They were also hoping that critical fragments of the spike protein would be sufficient to induce immunity, rather than the entire spike protein, if viral-like particles could be produced through augmentation with M and E (Lu et al., 2020).

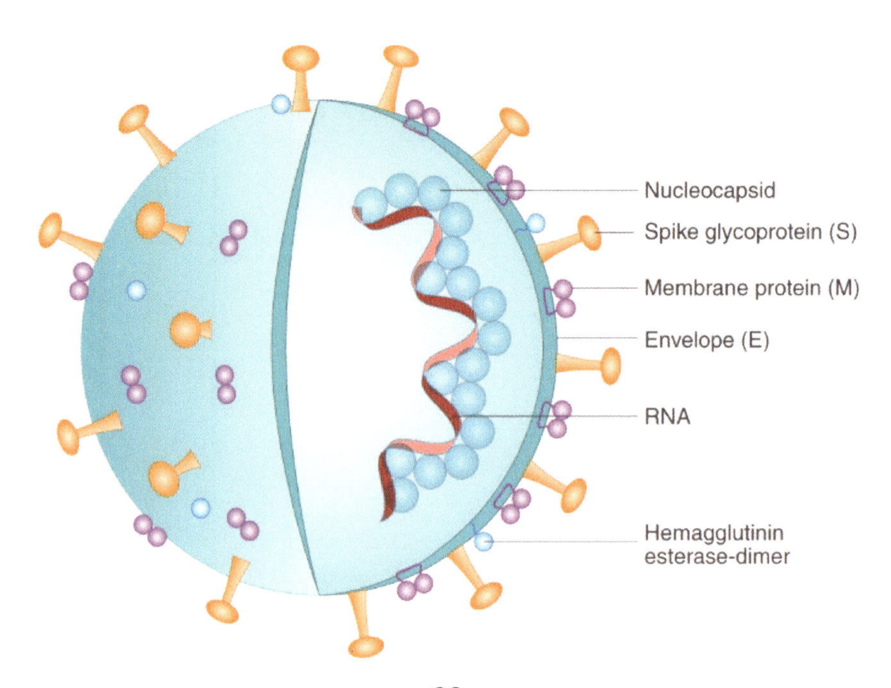

They confirmed experimentally that a vaccine containing the complete genes for all three proteins elicited a robust immune response that lasted for at least eight weeks following the second dose of the vaccine. Its performance was far superior to that of a vaccine containing only the spike protein. Disappointingly, a vaccine that contained only critical components of the spike protein, augmented with the other two envelope proteins, (structural proteins M and E) elicited practically no response.

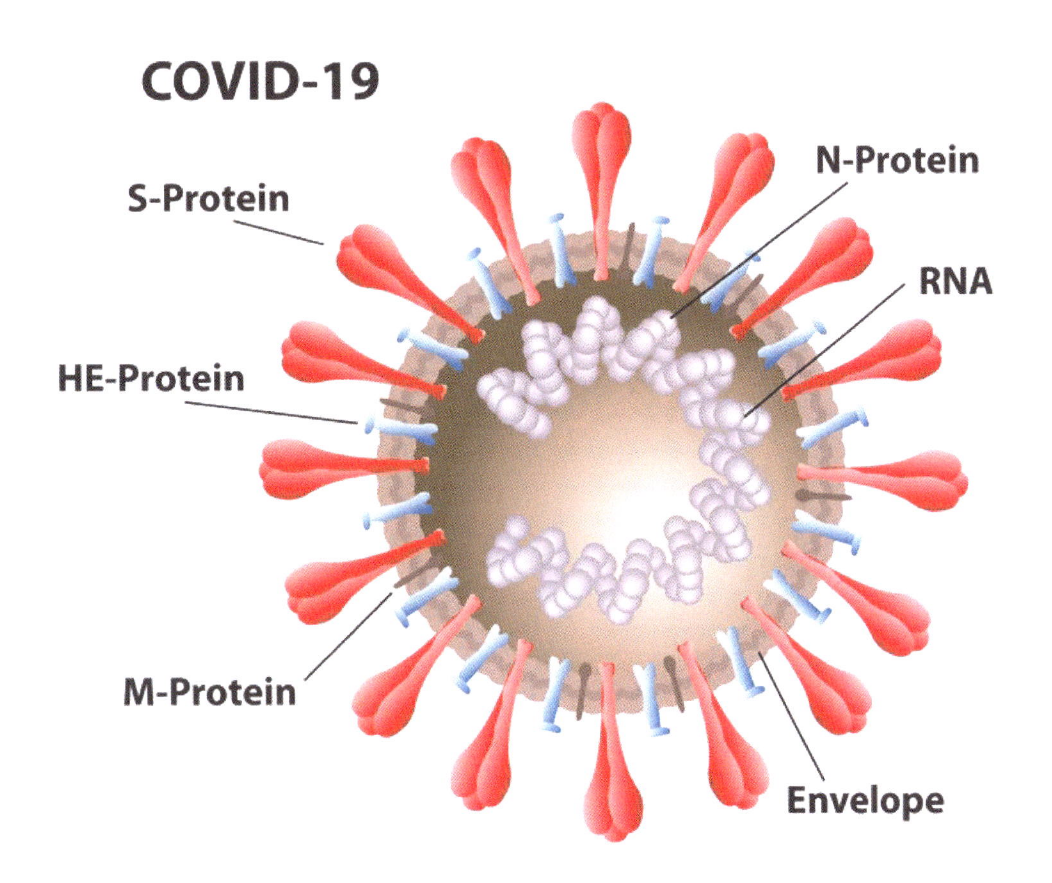

Moderna researchers have conducted similar studies with similar results. ***They concluded that the spike protein alone was clearly inferior to a formulation containing RNA encoding all three envelope proteins, (structural proteins M and E)*** **(Emphasis added) and they hypothesized that this was due to the fact that all three proteins were needed to allow the cell to release intact virus-like particles, rather than to just post the spike protein in the plasma membrane. The spike protein alone failed to initiate a T cell response in animal studies, whereas the formulation with all three (structural) proteins did (Corbett et al., 2020).**

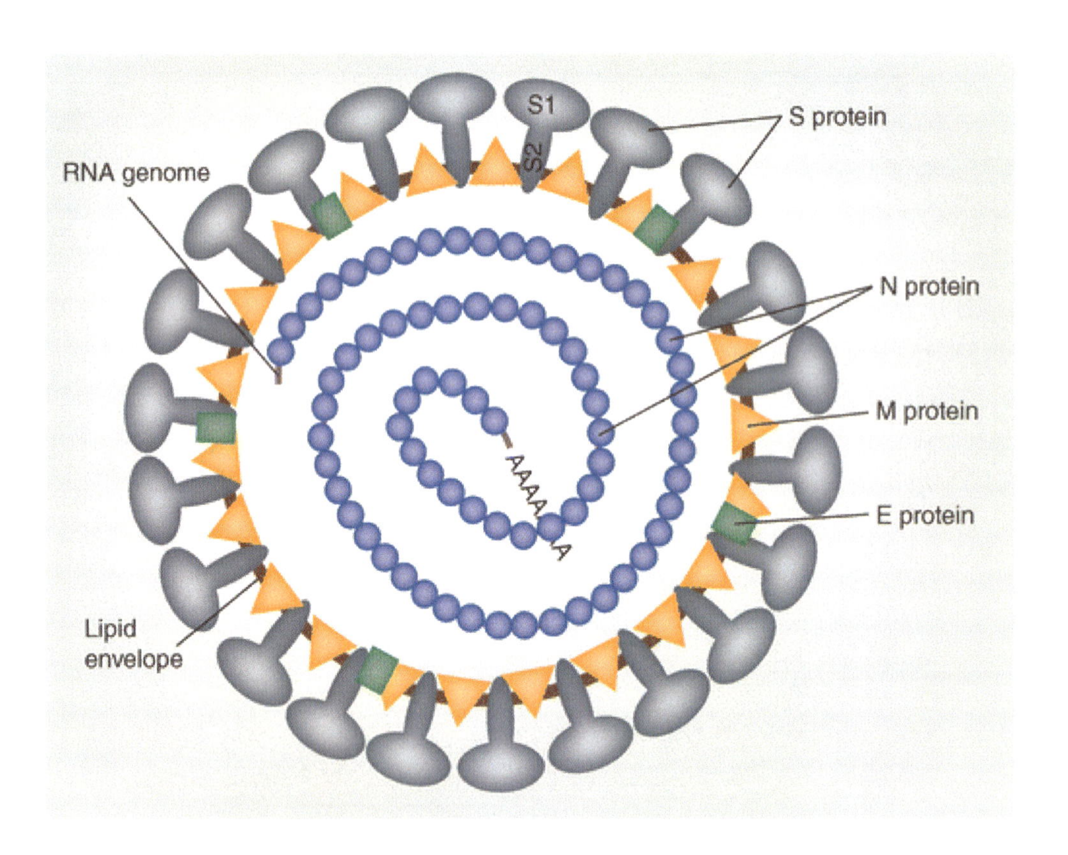

The two emergency-approved vaccines only contain mRNA code for spike protein **(Emphasis added) (without E or M), and there must have been a good reason for this decision, despite its observed poor performance. It is possible that more sophisticated design of the lipid nanoparticle resulted in the ability to have the lipids serve as an adjuvant (similar to aluminum that is commonly added to traditional vaccines) while still protecting the RNA from degradation.**

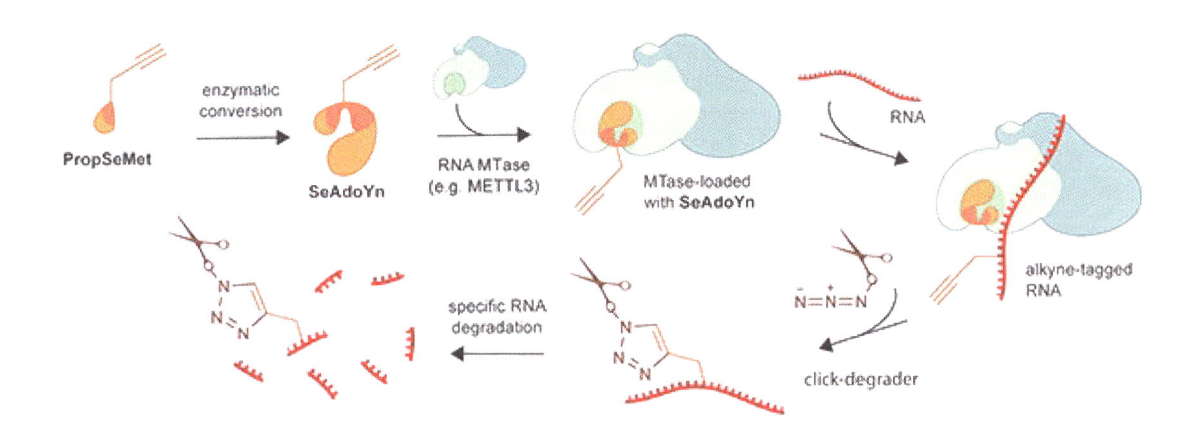

RNA degradation

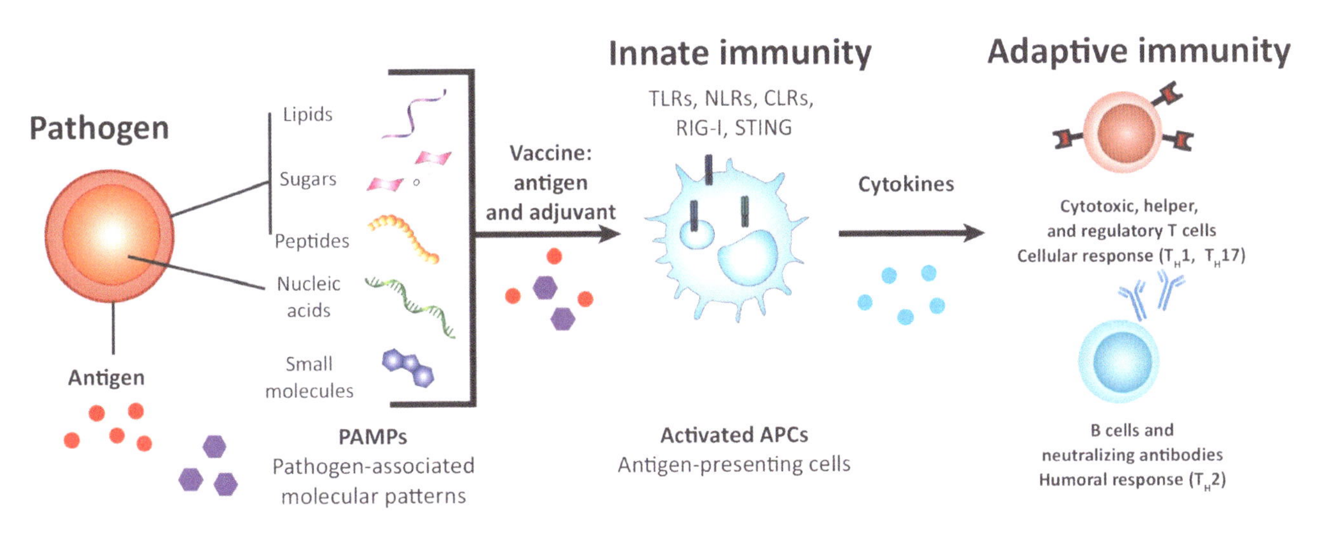

Trends in Biotechnology

Another curious modification in the RNA code is that the developers have enriched the sequence in cytosines and guanines (Cs and Gs) at the expense of adenines and uracils (As and Us). They have been careful to replace only the third position in the codon in this way, and only when it does not alter the amino acid map (Hubert, 2020). It has been demonstrated experimentally that *GC-rich mRNA sequences are expressed (translated into protein) (ie. spike protein) up to 100-fold more efficiently than GCpoor sequences* (Kudla et al., 2006). *So this appears to be another modification to further assure synthesis of abundant copies of the spike protein. We do not know the unintended consequences of this maneuver.* (Emphasis added)

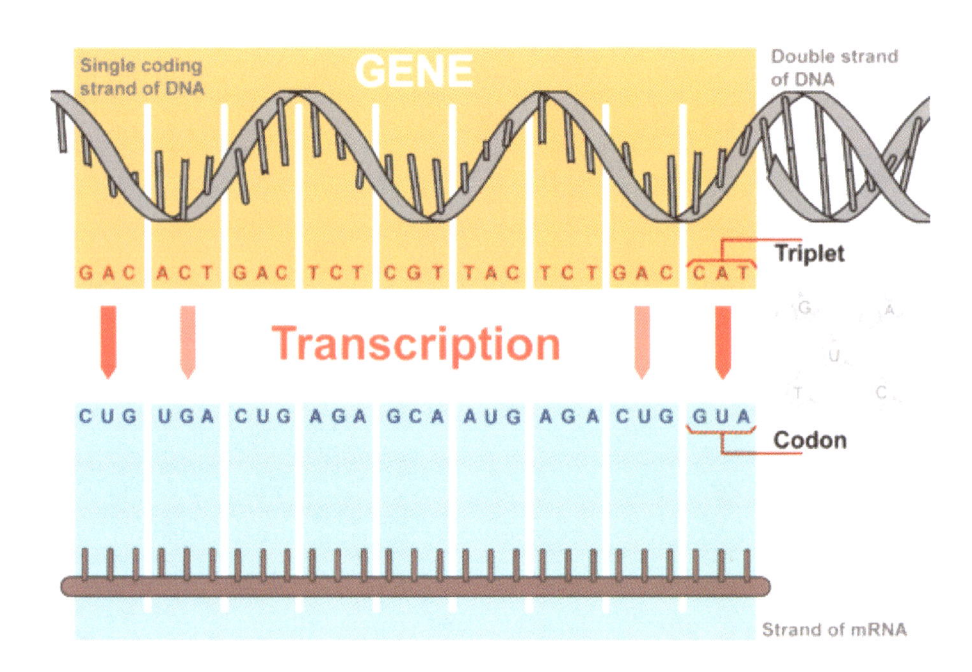

Intracellular pathogens, including viruses, tend to have low GC content compared to the host cell's genome (Rocha and Danchin, 2020). So, this modification may have been motivated in part by the desire to enhance the effectiveness of the deception that the protein is a human protein.

All of these various modifications to the RNA are designed to make it resist breakdown, appear more like a human messenger RNA protein-coding sequence, and efficiently translate into antigenic protein.

In April 2020 an important paper was published regarding the potential for self-reactive antibodies to be generated following exposure to the spike protein and other antigenic epitopes spread over the length of SARS-CoV-2. Lyons-Weiler (2020) coined the phrase "pathogen priming" because he believed the more commonly used "immune enhancement" fails to capture the severity of the condition and its consequences.

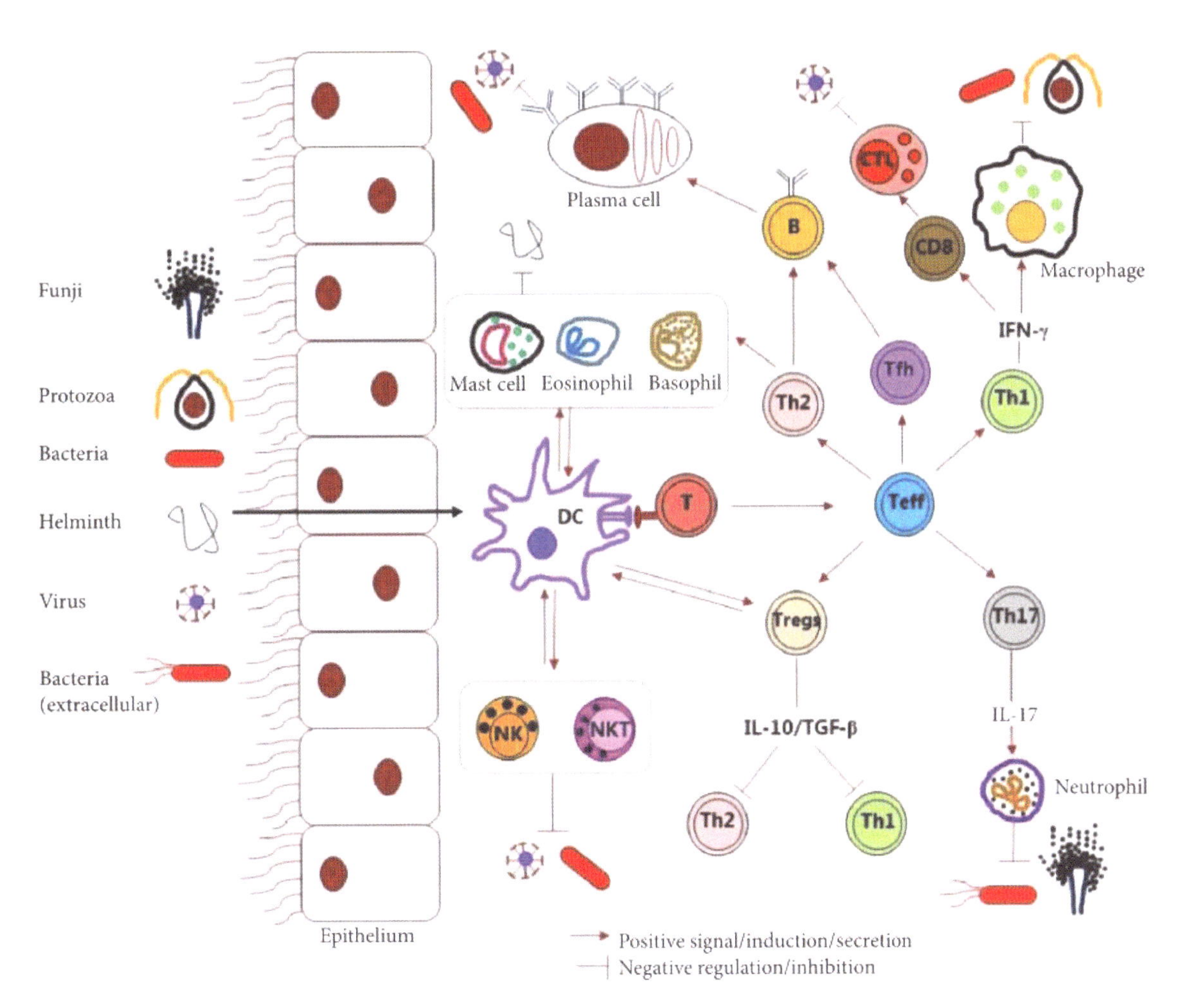

In his in silico analysis, Lyons-Weiler compared all antigenic SARSCoV-2 protein epitopes flagged in the SVMTriP database (http://sysbio.unl.edu/SVMTriP/) and searched the p-BLAST database (https://blast.ncbi.nlm.nih.gov/Blast.cgi) for homology between those epitopes and endogenous human proteins.

Of the 37 SARS-CoV-2 proteins analyzed, 29 had antigenic regions. All but one of these 29 had homology with human proteins (putative selfantigens) and were predicted to be autoreactogenic. The largest number of homologies were associated with the spike (S) protein and the NS3 protein, both having 6 homologous human proteins.

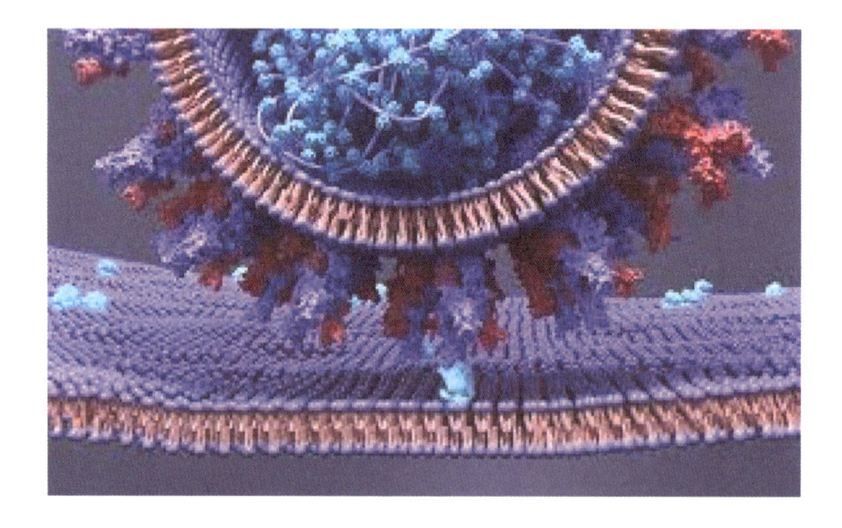

A functional analysis of the endogenous human proteins homologous with viral proteins found that over 1/3 of them are associated with the adaptive immune system. The author speculates that prior virus exposure or prior vaccination, either of which could initiate antibody production that targets these endogenous proteins, may be playing a role in the development of more severe disease in the elderly in particular. In this case the pre-existing antibodies act to suppress the adaptive immune system and lead to more severe disease.

Another group (Ehrenfeld et. al., 2020), in a paper predominantly about the wide range of autoimmune diseases found in association with a prior SARS-CoV-2 infection, also investigated how the spike protein could trigger such a range of diseases. They report, in Table 1 of that reference, strings of heptapeptides within the human proteome that overlap with the spike protein generated by SARS-CoV-2.

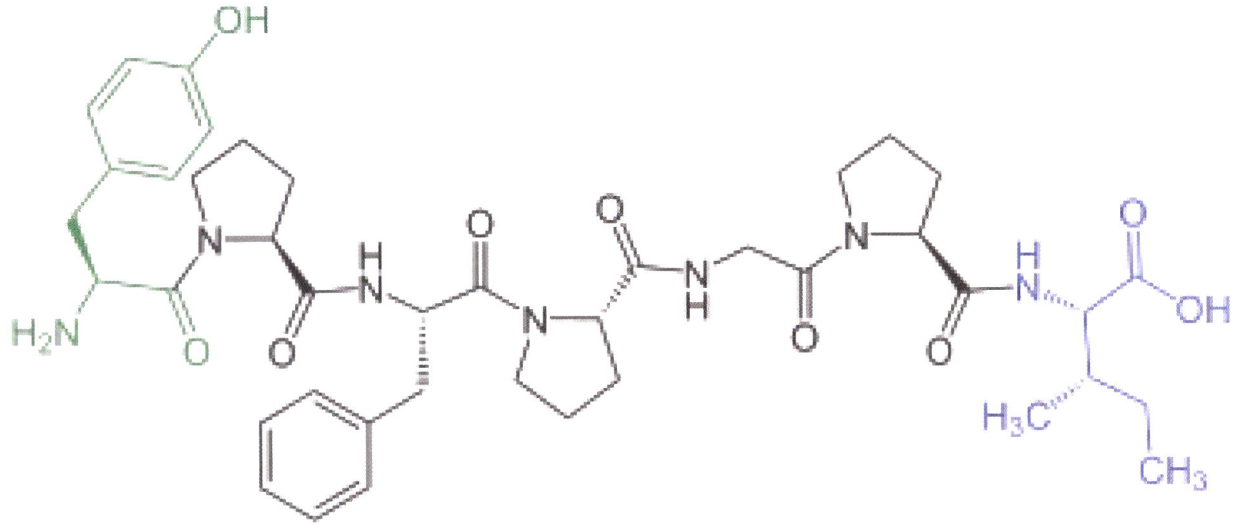

Heptapeptide

They identified 26 heptapeptides found in humans and in the spike protein. It is interesting to note that 2 of the 26 overlapping heptapeptides were found to be sequential, a strikingly long string of identical peptides to be found in common between endogenous human proteins and the spike protein. Commenting on the overlapping peptides they had discovered and the potential for this to drive many types of autoimmunity simultaneously, they comment, *"The clinical scenario that emerges is upsetting." Indeed, it is.* (Emphasis added)

In May of 2020 another important paper in this regard was published by Vojdani and Kharrazian (2020). The authors used both mouse and rabbit monoclonal antibodies against the 2003 SARS spike protein to test for reactivity against not only the spike protein of SARS-CoV-2, but also against several endogenous human proteins. They discovered that there was a high level of binding not only with the SARS-CoV-2 spike protein, but against a wide range of endogenous proteins. "[W]e found that the strongest reactions were with transglutaminase 3 (tTG3), transglutaminase 2 (tTG2), ENA, myelin basic protein (MBP), mitochondria, nuclear antigen (NA), α-myosin, thyroid peroxidase (TPO), collagen, claudin 5+6, and S100B." (Vojdani and Kharrazian, 2020).

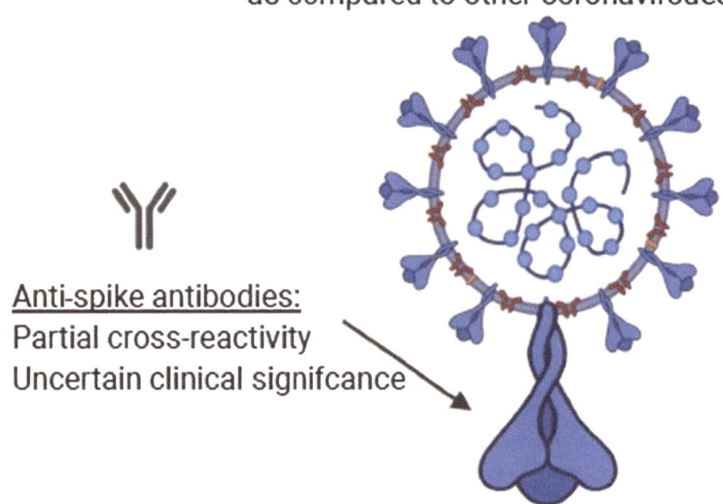

These important findings need to be emphasized. Antibodies with a high binding affinity to SARSCoV-2 spike and other proteins also have a high binding affinity with tTG (associated with Celiac Disease), TPO (Hashimoto's thyroiditis), myelin basic protein (multiple sclerosis), and several endogenous proteins. Unlike the autoimmune process associated with pathogen priming, these autoimmune diseases typically take years to manifest symptomatically.

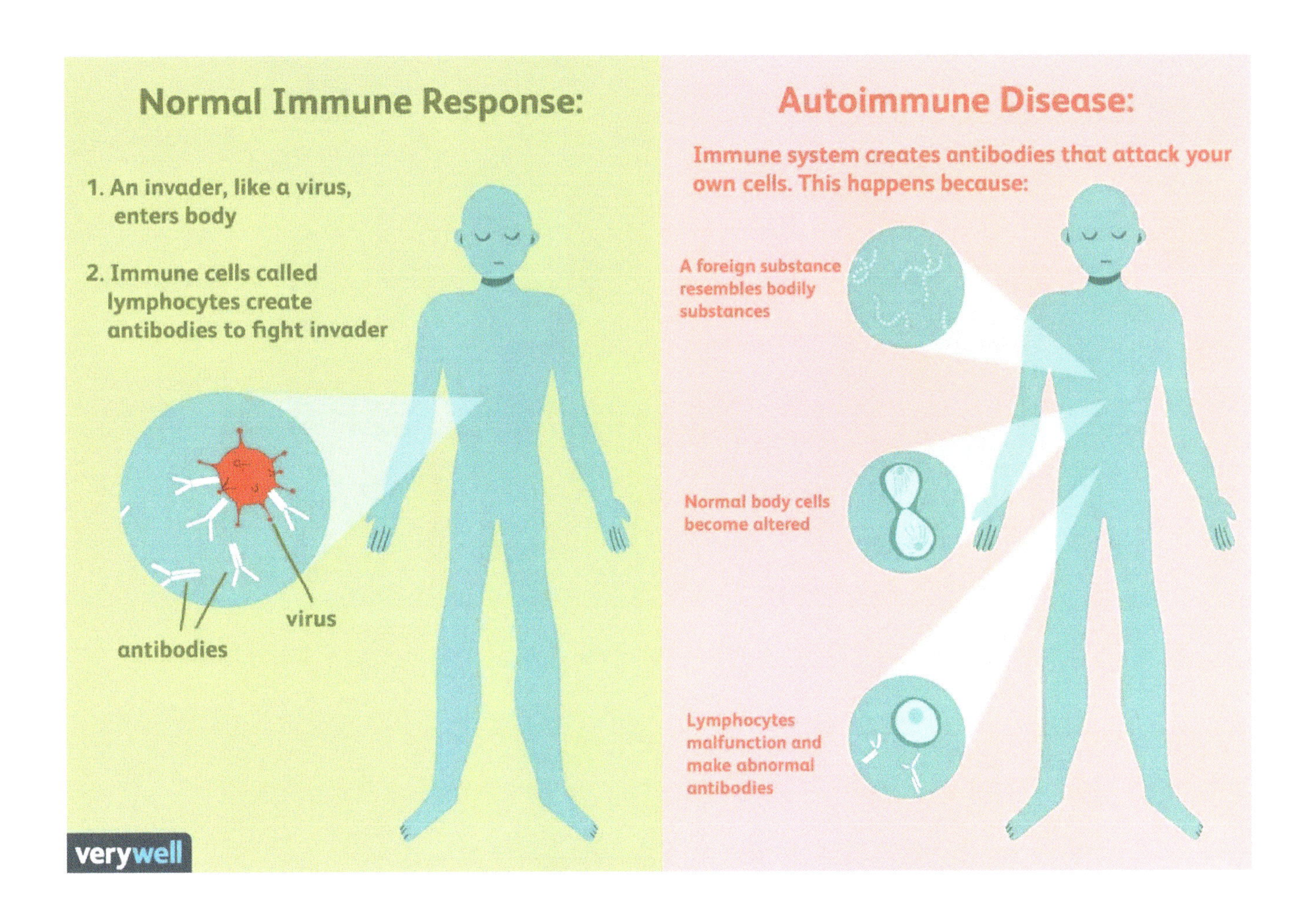

The autoantibodies generated by the spike protein predicted by Lyons-Weiler (2020) and described above were confirmed with an in vitro study published more recently. In this follow-on paper, Vojdani et. al., (2021) looked again at the issue of cross-reactivity of antibodies, this time using human monoclonal antibodies (mAbs) against the SARS-CoV-2 spike protein rather than mouse and rabbit mAbs. Their results confirmed and extended their prior findings. "At a cutoff of 0.32 OD [optical density], SARS-CoV-2 membrane protein antibody reacted with 18 out of the 55 tested antigens." These 18 endogenous antigens encompass reactivity to tissue in liver, mitochondria, the nervous and digestive system, the pancreas, and elsewhere in the body.

Human monoclonal antibodies (mAbs)

Autoimmunity is becoming much more widely recognized as a sequela of COVID-19. There are multiple reports of previously healthy individuals who developed diseases such as idiopathic thrombocytopenic purpura, Guillain-Barré syndrome and autoimmune haemolytic anaemia (Galeotti and Bayry, 2020). There are three independent case reports of systemic lupus erythemosus (SLE).

The same study, referencing the autoantibodies predicted by Lyons-Weiler (2020) mentioned above, notes with obvious grave concern: "The Sars-CoV-2 spike protein is a potential epitopic target for biomimicry-induced autoimmunological processes [25]. Therefore, we feel it will be extremely important to investigate whether GPCR-fAABs will also become detectable after immunisation by vaccination against the virus."

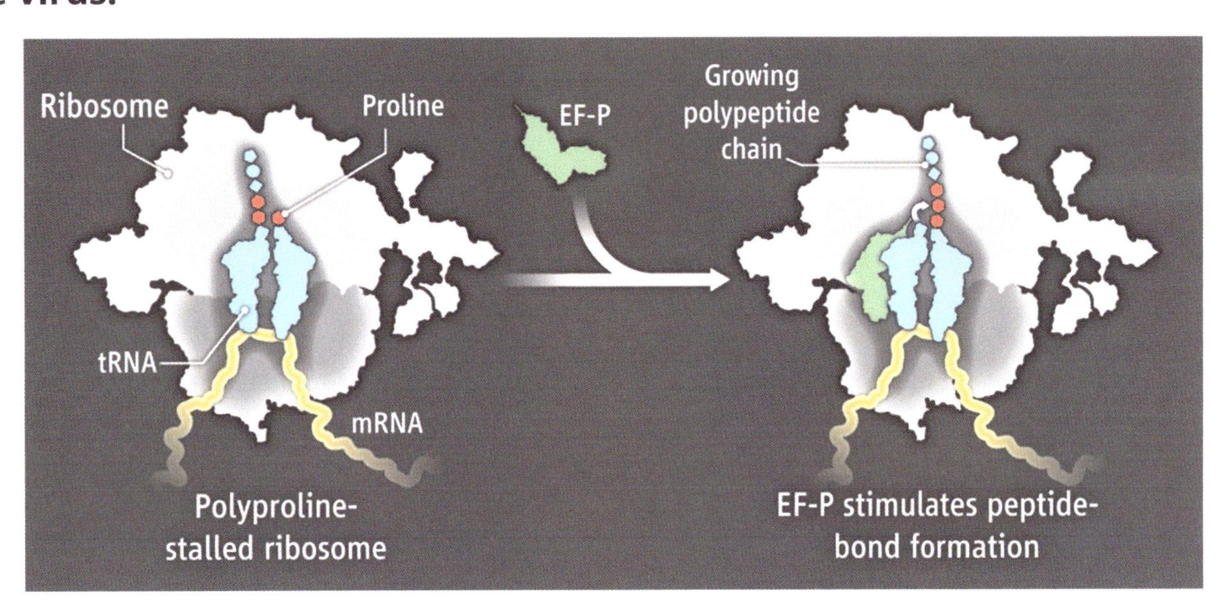

We have reviewed the evidence here that the spike protein of SARS-CoV-2 has extensive sequence homology with multiple endogenous human proteins and could prime the immune system toward development of both auto-inflammatory and autoimmune disease. This is particularly concerning given that the protein has been redesigned with two extra proline residues to potentially impede its clearance from the circulation through membrane fusion. These diseases could present acutely and over relatively short timespans such as with MIS-C or could potentially not manifest for months or years following exposure to the spike protein, whether via natural infection or via vaccination.

Many who test positive for COVID-19 express no symptoms. The number of asymptomatic, PCR positive cases varies widely between studies, from a low of 1.6% to a high of 56.5% (Gao et. al., 2020). Those who are insensitive to COVID-19 probably have a very strong innate immune system. The healthy mucosal barrier's neutrophils and macrophages rapidly clear the viruses, often without the need for any antibodies to be produced by the adaptive system. However, the vaccine intentionally completely bypasses the mucosal immune system, both through its injection past the natural mucosal barriers and its artificial configuration as an RNA-containing nanoparticle.

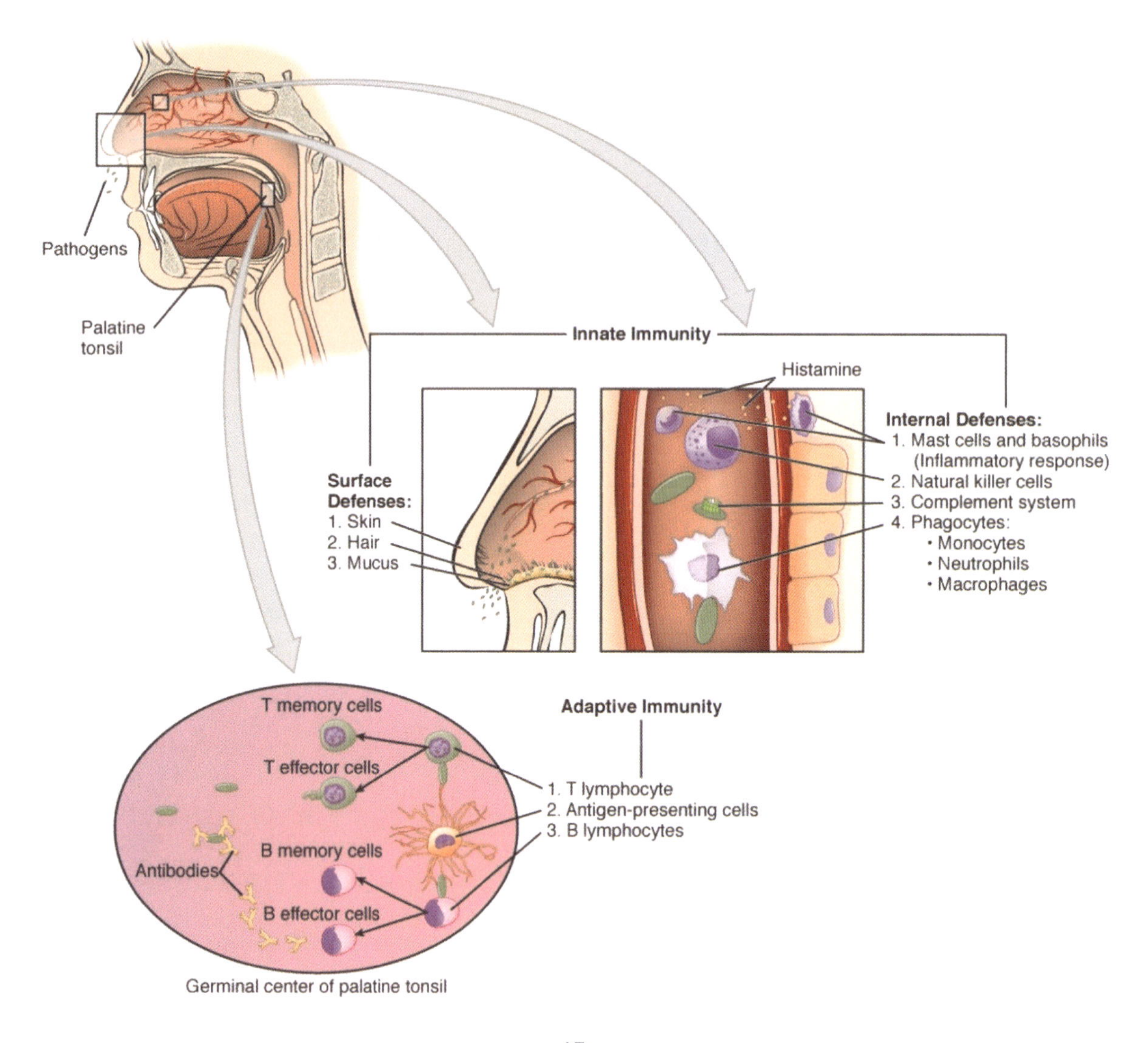

As noted in Carsetti (2020), those with a strong innate immune response almost universally experience either asymptomatic infection or only mild COVID-19 disease presentation. Nevertheless, they might face chronic autoimmune disease, as described previously, *as a consequence of excessive antibody production in response to the vaccine, which was not necessary in the first place.* (Emphasis added)

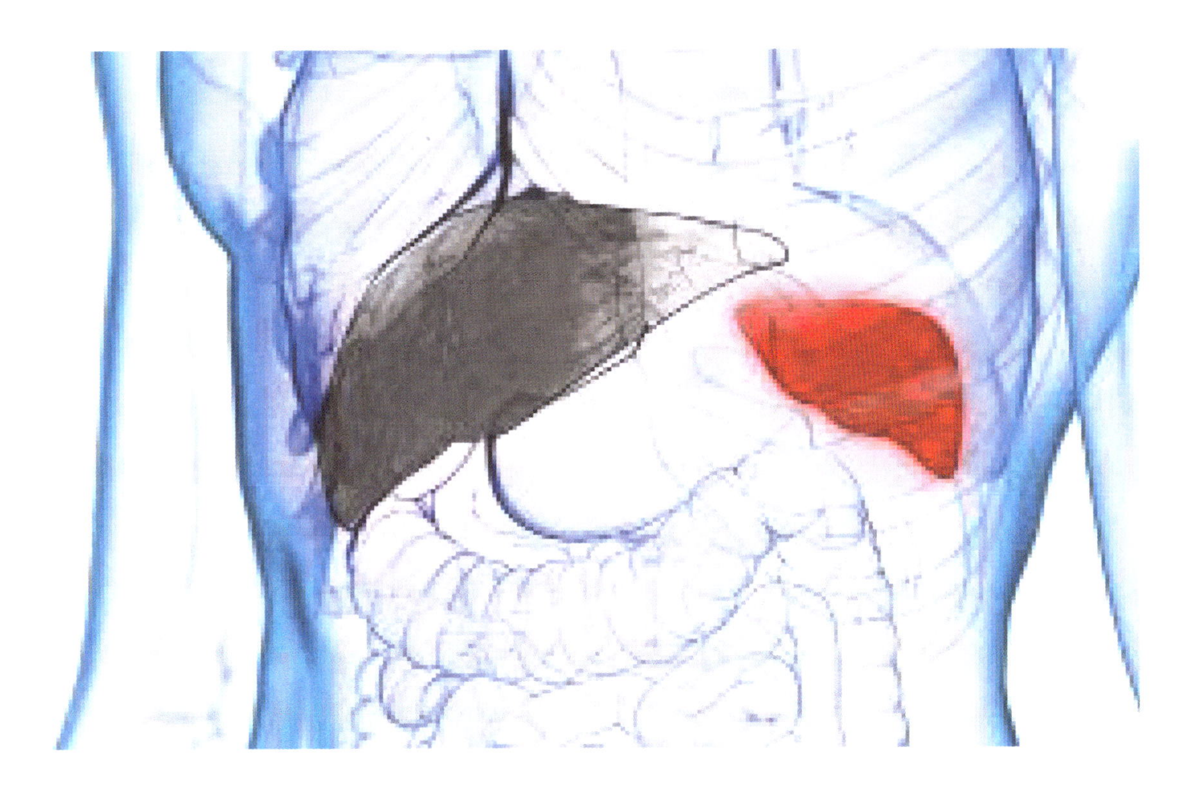

Several studies on mRNA-based vaccines have confirmed independently that the spleen is a major center of activity for the immune response. A study on an mRNA-based influenza virus vaccine is extremely relevant for answering the question of the biodistribution of the mRNA in the vaccine. This vaccine, like the SARS-CoV-2 vaccines, was designed as lipid nanoparticles with modified RNA coding for hemagglutinin (the equivalent surface fusion protein to the spike protein in corona viruses), and was administered through muscular injection.

They concluded that the mRNA distributes from the injection site to the liver and spleen via the lymphatic system, ultimately reaching the general circulation. This likely happens through its transport inside macrophages (white blood cells) and other immune cells that take it up at the muscular injection site. *Disturbingly, it also reaches into the brain, although at much lower levels* (Emphasis added) (Bahl et al., 2017). The European Medicines Agency assessment report for the Moderna vaccine also noted that mRNA could be detected in the brain following intramuscular administration at about 2% of the level found in the plasma (European Medicines Agency, 2021).

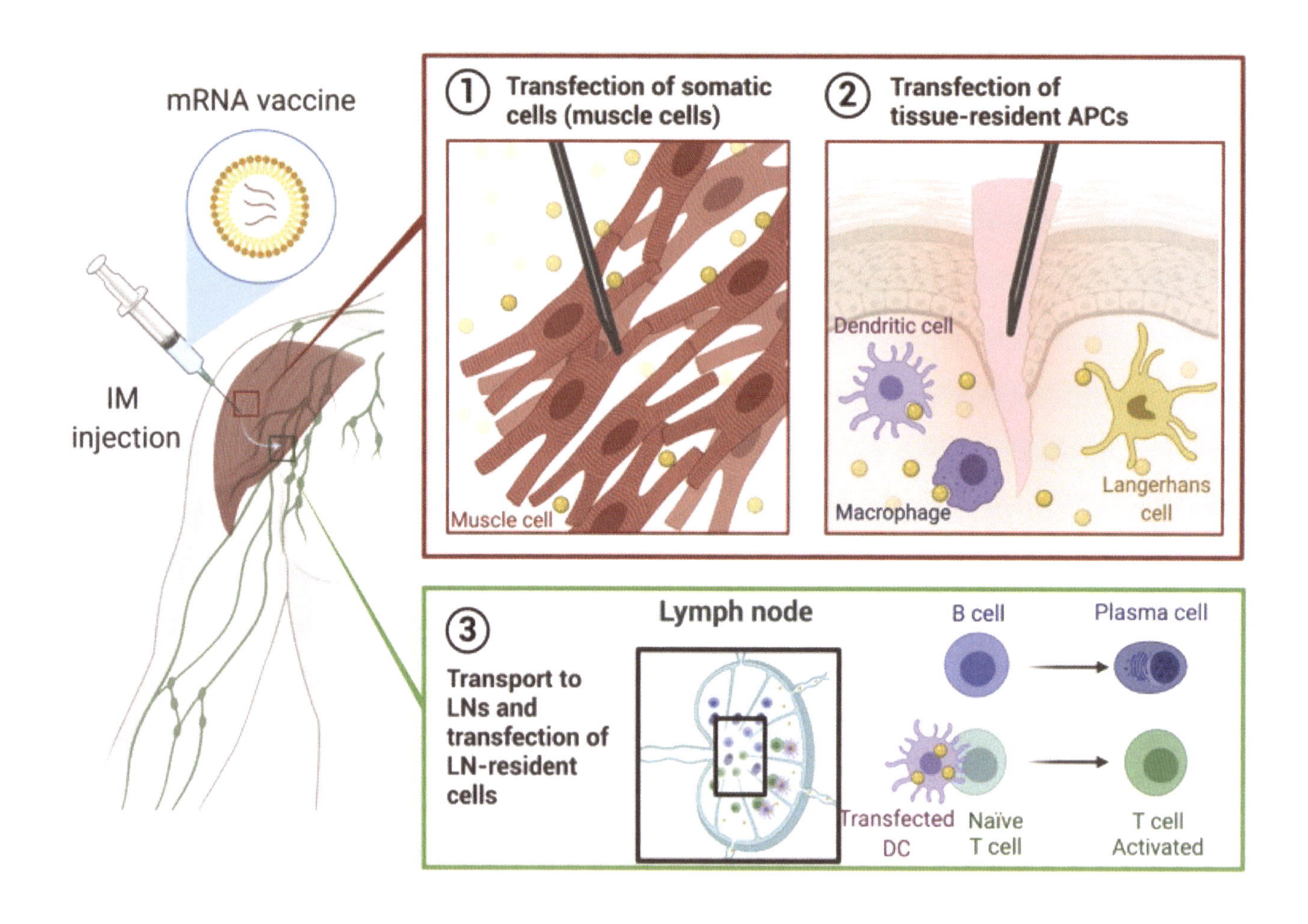

Finally, a study comparing luciferase-expressing mRNA nanoparticles with luciferase-expressing mRNA dendritic (antigen-presenting cells, APCs) cells as an alternative approach to vaccination revealed that the luciferase signal reached a broader range of lymphoid sites with the nanoparticle delivery mechanism. More importantly, the luciferase signal was concentrated in the spleen for the nanoparticles compared to dominance in the lungs for the dendritic cells (Firdessa-Fite and Creuso, 2020).

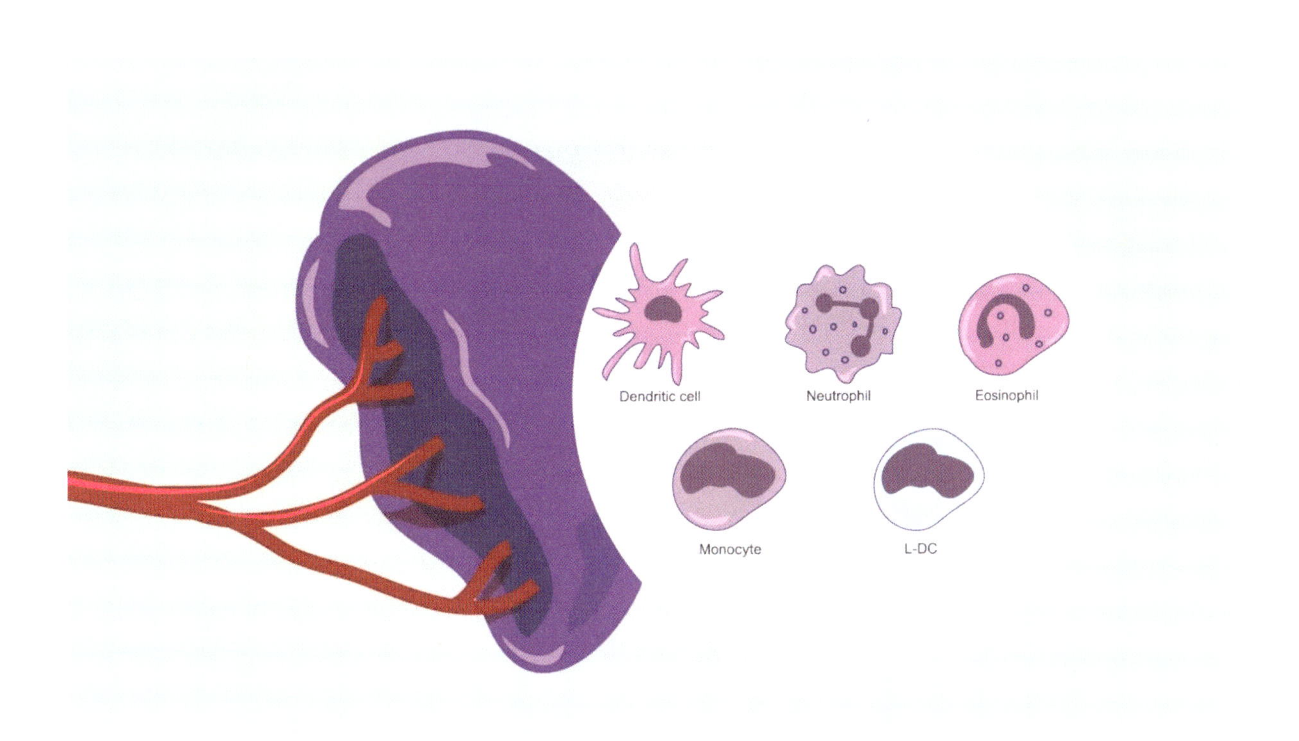

Thus, it seems plausible that a dangerous cascade leading to Immune thrombocytopenia (ITP) could ensue following mRNA vaccination, even with no live virus present, particularly in the context of impaired autophagy. Immune cells in the arm muscle take up the RNA particles and circulate within the lymph system, accumulating in the spleen. There, the immune cells produce abundant spike protein, which binds to the platelet glycoproteins and desialylates them. Platelet interaction with neutrophils causes NETosis and the launch of an inflammatory cascade. The exposed glycoproteins become targets for autoimmune antibodies that then attack and remove the platelets, leading to a rapid drop in platelet counts, and a life-threatening event.

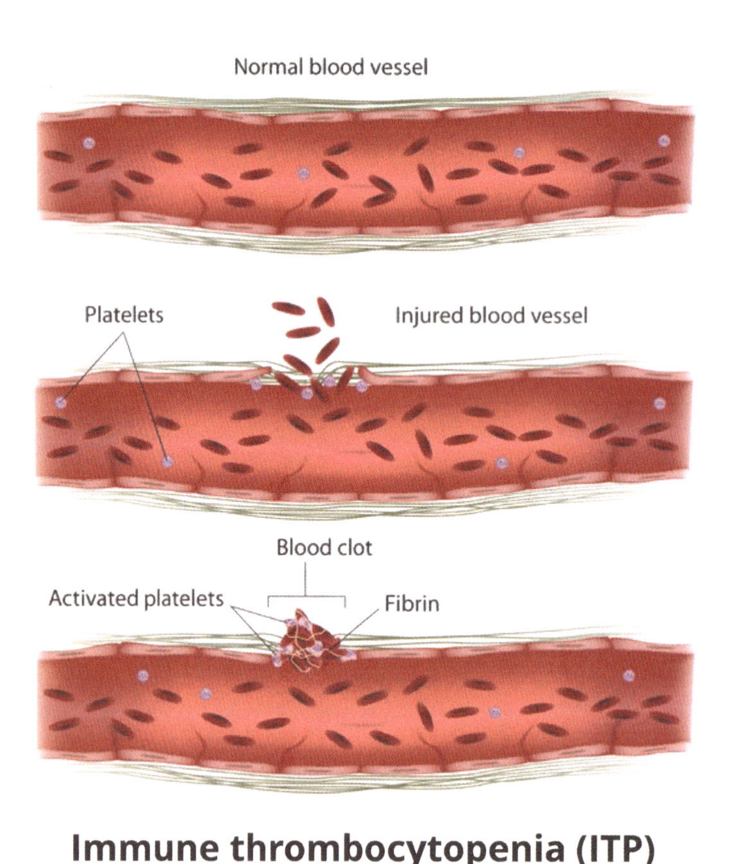

Immune thrombocytopenia (ITP)

The picture is now emerging that SARS-CoV-2 has serious effects on the vasculature in multiple organs, including the brain vasculature. As mentioned earlier, the spike protein facilitates entry of the virus into a host cell by binding to ACE2 in the plasma membrane. ACE2 is a type I integral membrane protein that cleaves angiotensin II into angiotensin(1-7), thus clearing angiotensin II and lowering blood pressure. In a series of papers, Yuichiro Suzuki in collaboration with other authors presented a strong argument that the spike protein by itself can cause a signaling response in the vasculature with potentially widespread consequences (Suzuki, 2020; Suzuki et al., 2020; Suzuki et al., 2021; Suzuki and Gychka, 2021).

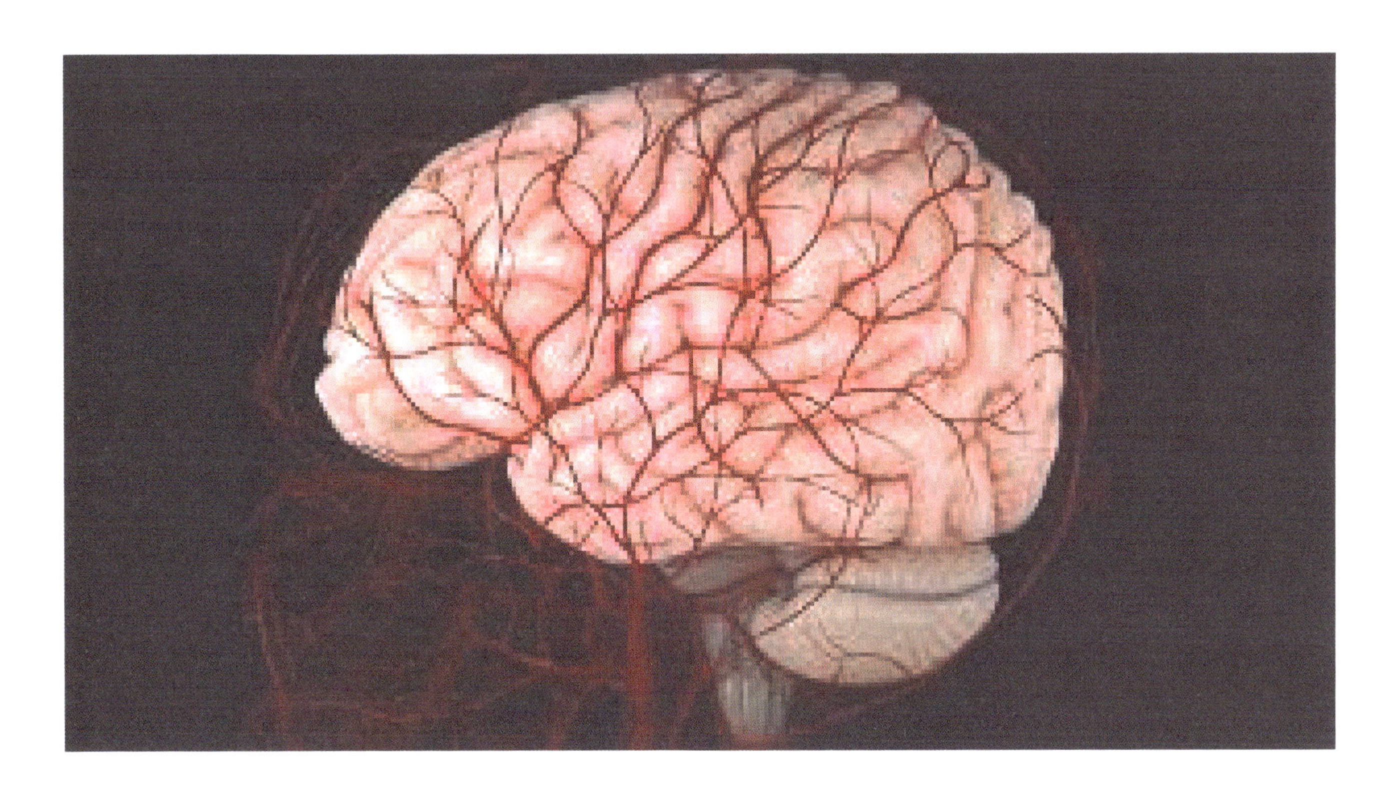

These authors observed that, in severe cases of COVID-19, SARSCoV-2 causes significant morphological changes to the pulmonary vasculature. Post-mortem analysis of the lungs of patients who died from COVID-19 revealed histological features showing vascular wall thickening, mainly due to hypertrophy of the tunica media. Enlarged smooth muscle cells had become rounded, with swollen nuclei and cytoplasmic vacuoles (Suzuki et al., 2020). Furthermore, they showed that exposure of cultured human pulmonary artery smooth muscle cells to the SARSCoV-2 spike protein S1 subunit was sufficient to promote cell signaling without the rest of the virus components.

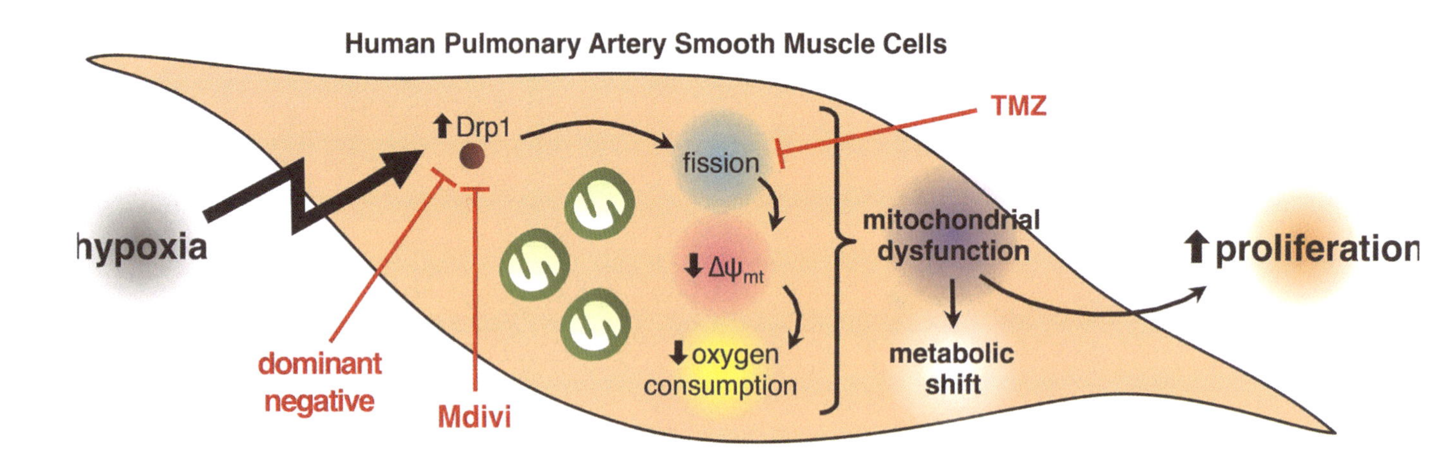

Follow-on papers (Suzuki et al., 2021, Suzuki and Gychka, 2021) showed that the spike protein S1 subunit suppresses ACE2, causing a condition resembling pulmonary arterial hypertension (PAH), a severe lung disease with very high mortality. Their model is depicted here in Figure 2. Ominously, Suzuki and Gychka (2021) wrote: "Thus, these in vivo studies demonstrated that the spike protein of SARS-CoV-1 (without the rest of the virus) reduces the ACE2 expression, increases the level of angiotensin II, and exacerbates the lung injury."

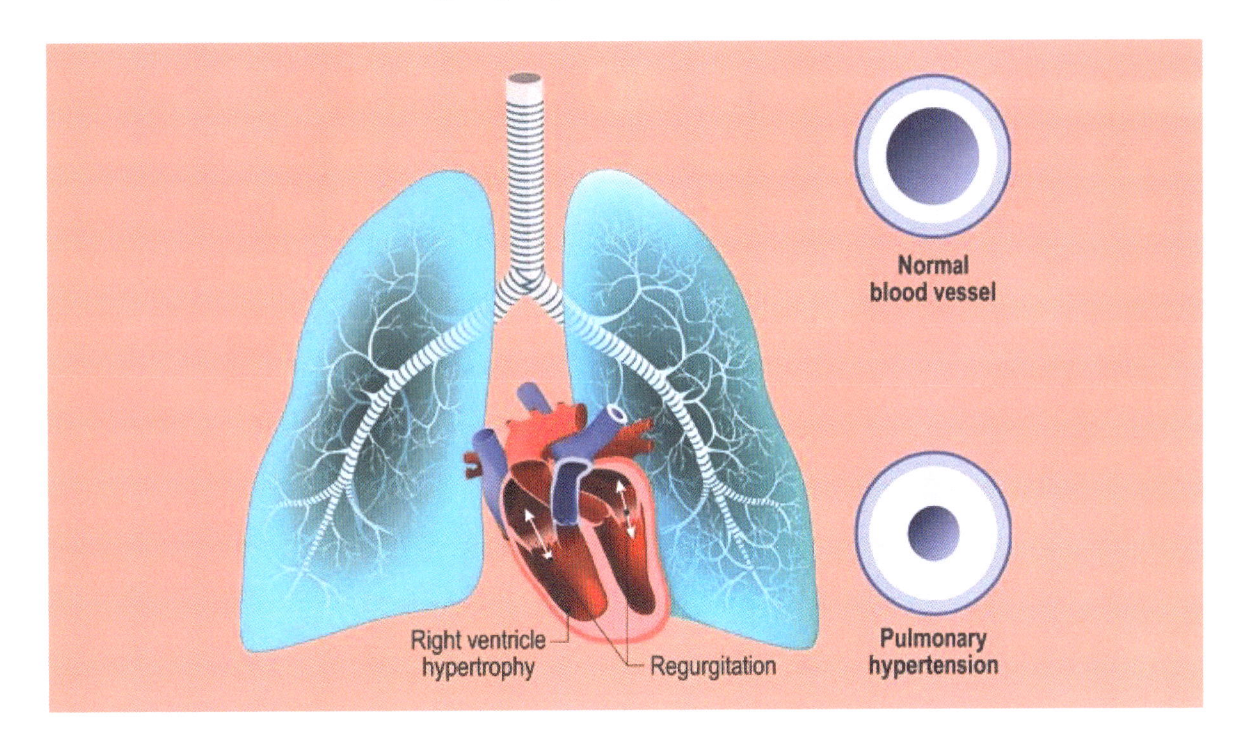

The "in vivo studies" they referred to here (Kuba et al., 2005) had shown that SARS coronavirus-induced lung injury was primarily due to inhibition of ACE2 by the SARS-CoV spike protein, causing a large increase in angiotensin-II. Suzuki et al. (2021) went on to demonstrate experimentally that the S1 component of the SARS-CoV-2 virus, at a low concentration of 130 pM, activated the MEK/ERK/MAPK signaling pathway to promote cell growth. They speculated that these effects would not be restricted to the lung vasculature.

The signaling cascade triggered in the heart vasculature would cause coronary artery disease, and activation in the brain could lead to stroke. Systemic hypertension would also be predicted. They hypothesized that this ability of the spike protein to promote pulmonary arterial hypertension could predispose patients who recover from SARS-CoV-2 to later develop right ventricular heart failure.

Figure 2: Mechanisms of Right Ventricular Dysfunction

Altered rhythm
Increased afterload
Abnormal preload
RA
RV
Lungs
LA
LV
Altered interdependence
Reduced contractility

LA = left atrium; LV = left ventricle; RA = right atrium; RV = right ventricle.

Furthermore, they suggested that a similar effect could happen in response to the mRNA vaccines, and they warned of potential longterm consequences to both children and adults who received COVID-19 vaccines based on the spike protein (Suzuki and Gychka, 2021).

An interesting study by Lei et. al. (2021) found that pseudovirus — spheres decorated with the SARS-CoV-2 S1 protein but lacking any viral DNA in their core — caused inflammation and damage in both the arteries and lungs of mice exposed intratracheally. They then exposed healthy human endothelial cells to the same pseudovirus particles. Binding of these particles to endothelial ACE2 receptors led to mitochondrial damage and fragmentation in those endothelial cells, leading to the characteristic pathological changes in the associated tissue.

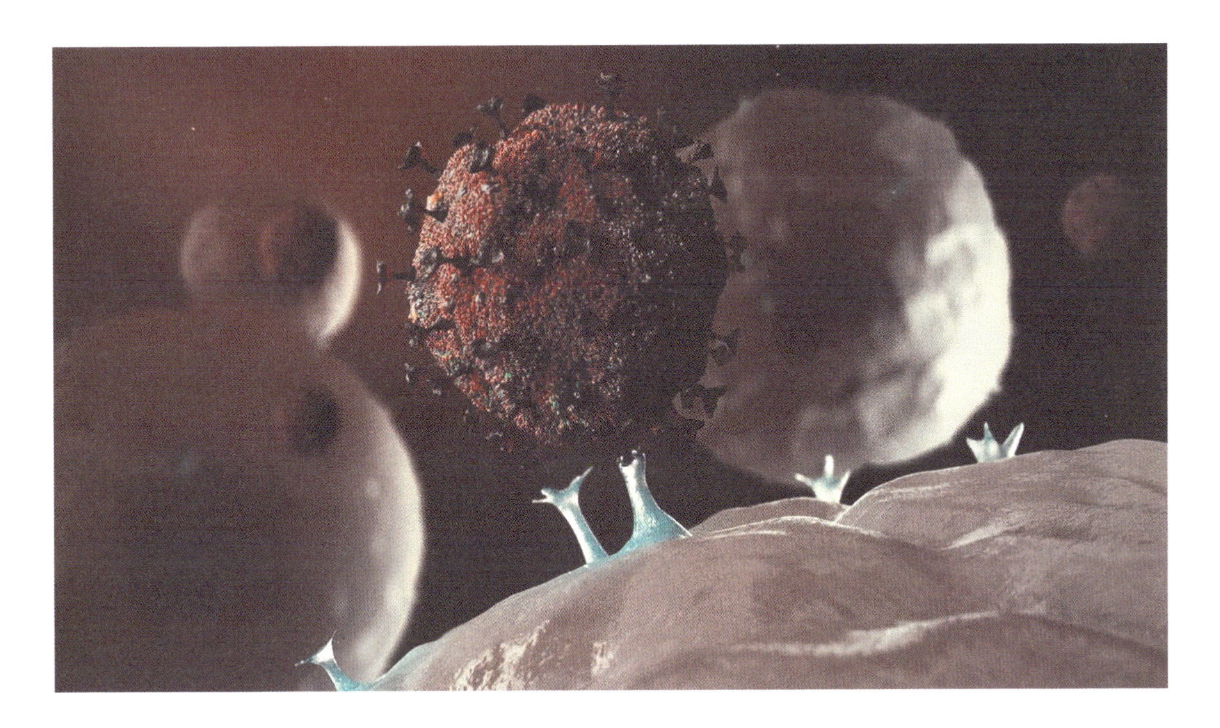

This study makes it clear that spike protein alone, unassociated with the rest of the viral genome, is sufficient to cause the endothelial damage associated with COVID-19. The implications for vaccines intended to cause cells to manufacture the spike protein are clear and are an obvious cause for concern.

Neurological symptoms associated with COVID-19, such as headache, nausea and dizziness, encephalitis and fatal brain blood clots are all indicators of damaging viral effects on the brain. Buzhdygan et al. (2020) proposed that primary human brain microvascular endothelial cells could cause these symptoms. ACE2 is ubiquitously expressed in the endothelial cells in the brain capillaries. ACE2 expression is upregulated in the brain vasculature in association with dementia and hypertension, both of which are risk factors for bad outcomes from COVID-19.

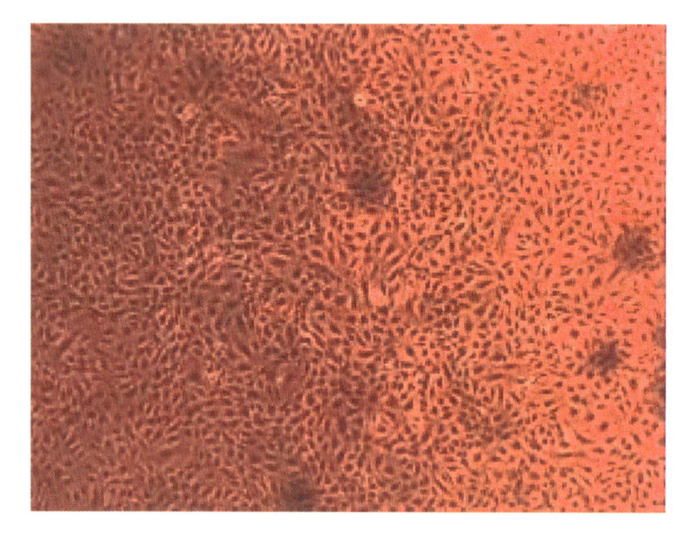

Human brain microvascular endothelial cells

In an in vitro study of the blood-brain barrier, the S1 component of the spike protein promoted loss of barrier integrity, suggesting that the spike protein acting alone triggers a pro-inflammatory response in brain endothelial cells, which could explain the neurological consequences of the disease (Buzhdygan et al., 2020). The implications of this observation are disturbing because the mRNA vaccines induce synthesis of the spike protein, which could theoretically act in a similar way to harm the brain.

The spike protein generated endogenously by the vaccine could also negatively impact the male testes, as the ACE2 receptor is highly expressed in Leydig cells in the testes (Verma et al., 2020). Several studies have now shown that the coronavirus spike protein is able to gain access to cells in the testes via the ACE2 receptor, and disrupt male reproduction (Navarra et al., 2020; Wang and Xu, 2020). A paper involving postmortem examination of testicles of six male COVID-19 patients found microscopic evidence of spike protein in interstitial cells in the testes of patients with damaged testicles (Achua et al., 2021).

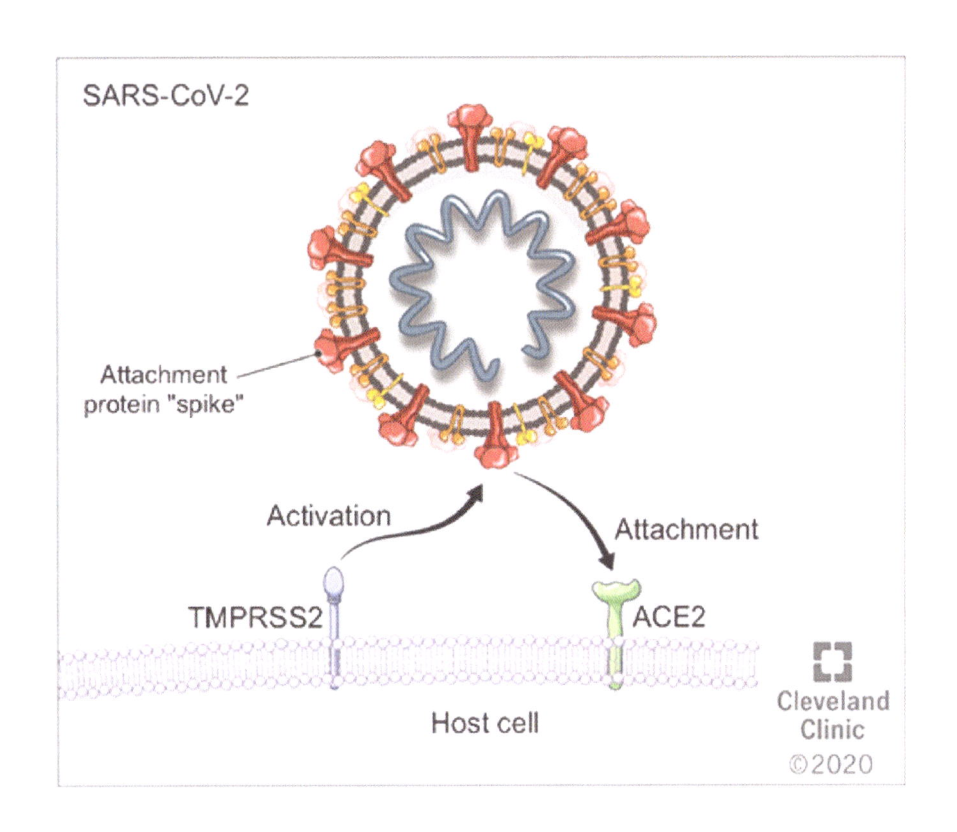

There has been considerable chatter on the Internet about the possibility of vaccinated people causing disease in unvaccinated people in close proximity. While this may seem hard to believe, there is a plausible process by which it could occur through the release of exosomes from dendritic cells in the spleen containing misfolded spike proteins, in complex with other prion reconformed proteins. These exosomes can travel to distant places. It is not impossible to imagine that they are being released from the lungs and inhaled by a nearby person. Extracellular vesicles, including exosomes, have been detected in sputum, mucus, epithelial lining fluid, and bronchoalveolar lavage fluid in association with respiratory diseases (Lucchetti et al., 2021)

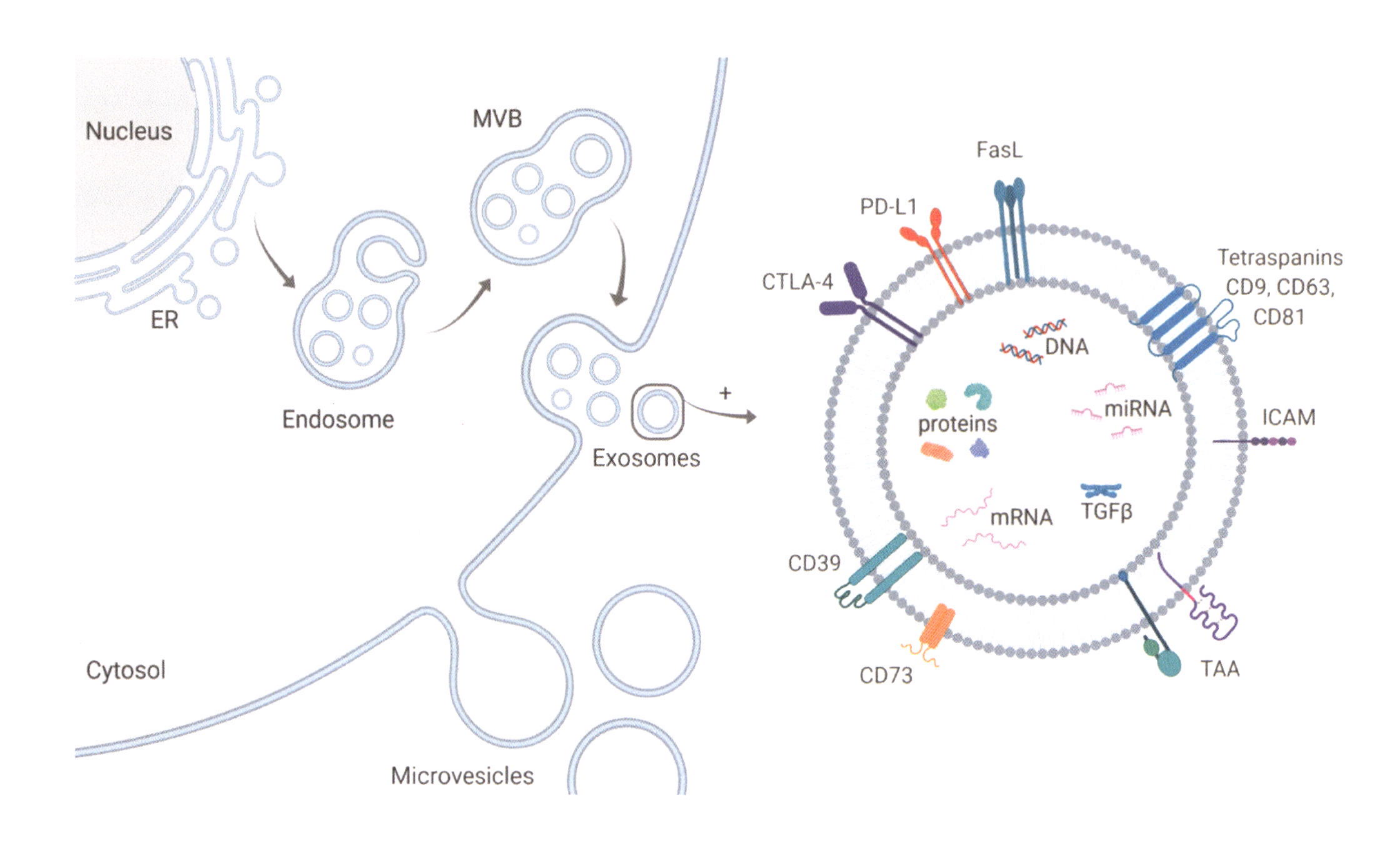

Phase 1/2/3 study undertaken by BioNTech on the Pfizer mRNA vaccine implied in their study protocol that they anticipated the possibility of secondary exposure to the vaccine (BioNTech, 2020). The protocol included the requirement that "exposure during pregnancy" should be reported by the study participants. They then gave examples of "environmental exposure during pregnancy" which included exposure "to the study intervention by inhalation or skin contact." They even suggested two levels of indirect exposure: "A male family member or healthcare provider who has been exposed to the study intervention by inhalation or skin contact then exposes his female partner prior to or around the time of conception."

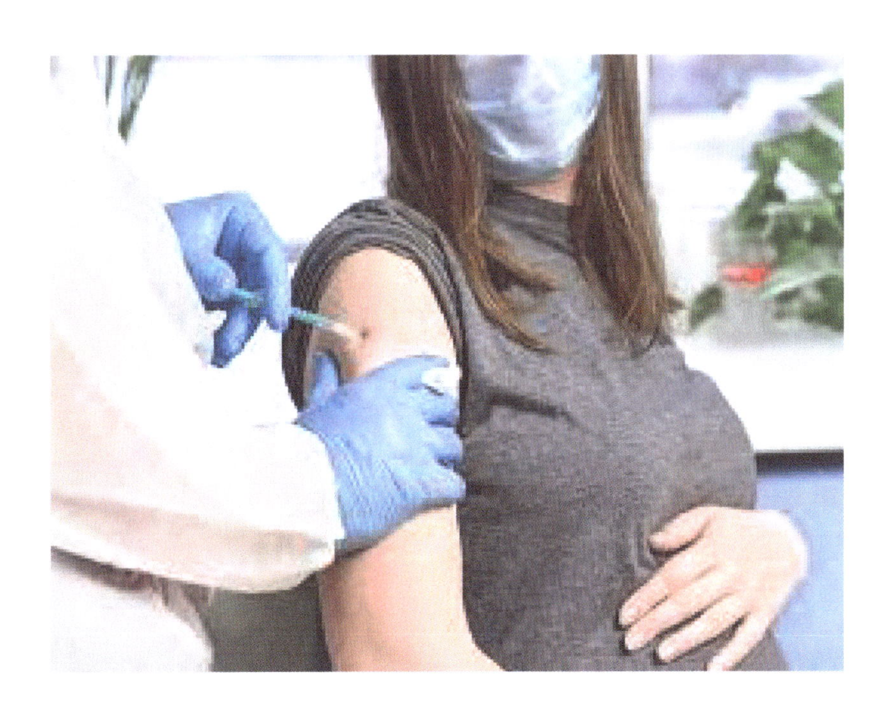

There are at least two concerns that we have regarding this experiment, in relation to the mRNA vaccines. The first is that, via continued infection of immune-compromised patients, we can expect continued emergence of more novel strains that are resistant to the antibodies induced by the vaccine, *such that the vaccine may quickly become obsolete, and there may well be demands for the population to undergo another mass vaccination campaign* (Emphasis added). Already a published study by researchers from Pfizer has shown that vaccine effectiveness is reduced for many of these variant strains. The vaccine was only 2/3 as effective against the South African strain as against the original strain (Liu et al., 2021).

Alpha Beta Gamma Delta

B.1.1.7 B.1.351 P.1 B.1.617.2

New names proposed for Covid variants

Country/region	Scientific name	WHO name
Kent, UK	B.1.1.7	Alpha
South Africa	B.1.351	Beta
Brazil	P.1	Gamma
India	B.1.617.2	Delta

Source: WHO BBC

The second more ominous consideration is to ponder what will happen with an immunecompromised patient following vaccination. It is conceivable that they will respond to the vaccine by producing antibodies, but those antibodies will be unable to contain the disease following exposure to COVID-19 due to impaired function of cytotoxic T cells. This scenario is not much different from the administration of convalescent plasma to immune-compromised patients, and so it might engender the evolution of antibody-resistant strains in the same way, only on a much grander scale. ***This possibility will surely be used to argue for repeated rounds of vaccines every few months, with increasing numbers of viral variants coded into the vaccines. This is an arms race that we will probably lose.*** **(Emphasis added)**

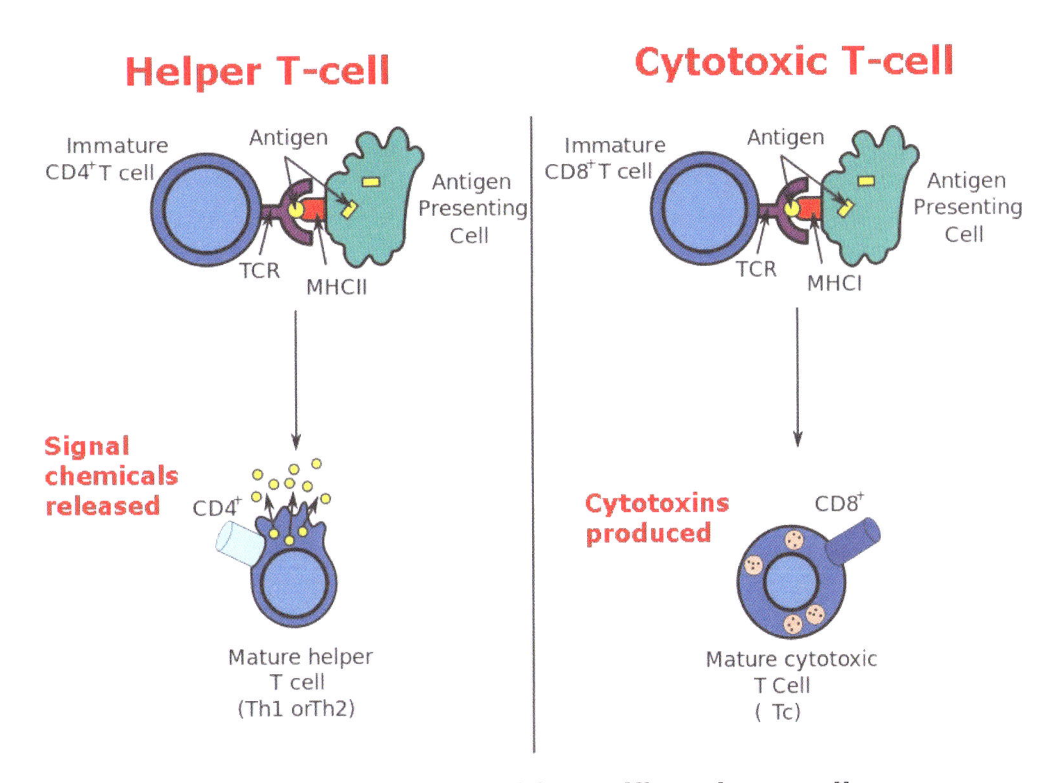

TCR = T-cell receptor (interacts with specific antigen on disease-causing cell or virus or infected human cell)

Revelation 6:2

2 So looked and saw a white horse, and its rider held a bow. And he was given a crown, and he rode out to overcome and conquer.

If I were to decipher this verse in light of current times, the bow and arrow, as a weapon, are likened to the modern day needle and syringe. To administer a vaccine, one must pull back on the plunger to draw in the liquid formula through the hollow needle (arrow/pointed objects/charagma). The contents then are directed at a human being and launched into the bloodstream, with the target being our cells.

The white horse and clothing are emblematic of white lab coats worn by doctors and scientists, whose diagnosis and treatment plans have the ability to administer life or death to an individual. According to Dictionary.com, the word "corona" entered the English language around 1555-65. It was borrowed directly from the Latin corōna, meaning "garland, wreath, crown."

The pandemic has served to overcome the global population and to conquer it in the form of the coming "great reset". Anthony Patch

For more information, please visit: https://www.anthonypatch.com

Printed in Great Britain
by Amazon

43759033R00037